Research in Biotechnology

Principles of Experimental Design in Biotechnology
Rock Canyon High School

Volume 2
April 2017

Editors
Shawndra L. Fordham
Wendy E. Lerolland
Bryan M. Winkelman

COVER PHOTO CREDITS

Front Cover:
Danio rerio embryos GFP. Alexis Chan, Alexi Brown, Tessa Rowe

Rear Cover *(left to right, top to bottom)*:
Plants. Tina Gilbert, Nathan Watervoort
Zebrafish embryo. CSIRO [CC BY 3.0 (http://creativecommons.org/licenses/by/3.0)], via Wikimedia Commons
Lion's mane mushroom. Melissa McMasters, Memphis, TN, US (Lion's mane mushroom) [CC BY 2.0 (http://creativecommons.org/licenses/by/2.0)], via Wikimedia Commons
Epithelial Tissue. Waverly Shannon, Mallika Kumar, and Sriya Sharma
C. elegans by Wormbase.org [used by permission] via WormClassroom.org
GFP *C. elegans*: Dan Dickinson, Goldstein lab, UNC Chapel Hill, http://wormcas9hr.weebly.com/ - Own work, [CC BY-SA 3.0], via Wikimedia Commons
Oil Bottle. Haley Swamberger, Ariel Lepard, Hannah Philip, Amy England
Hibiscus. Daniel Ramirez, Honolulu, USA [CC BY 2.0 (http://creativecommons.org/licenses/by/2.0)], via Wikimedia Commons
M. avium (mycobacteria). https://www.flickr.com/photos/pulmonary_pathology/7342846772
Columbine. Christine Majul. https://www.flickr.com/photos/kitkaphotogirl/3669786572 [CC BY 2.0 (http://creativecommons.org/licenses/by/2.0)]
3D Bioprinter. Sam Martin and Bailey Timmons
Drosophila melanogaster (VIII). Paco Romero-Ferrero.
https://www.flickr.com/photos/pacoromeroferrero/21527004263/[CC BY 2.0 (http://creativecommons.org/licenses/by/2.0)]
FoldScope (bottom left): Kyle Franklin, Colby Weintraub, and Ansh Jerath
Green Tea. McKay Savage from London, UK - China - Chengdu 4 - green tea in the park, [CC BY 2.0 (http://creativecommons.org/licenses/by/2.0)], via Wikimedia Commons

Cover designed by Shawndra L. Fordham

ACKNOWLEDGMENTS

The success of the second year of Rock Canyon High School's Principles of Experimental Design in Biotechnology has been the result of a combined effort of so many people, from our generous research funders, to our amazing mentors, and everyone in between. First I would like to thank Wendy Lerolland for partnering with me this year and teaching the technical writing course that all the research students took. Wendy was an invaluable resource to the students and me and helped to grow this program and improve on it in every aspect. I am fortunate to have had the opportunity to work with her. I also want to extend a special thank you to Bryan Winkelman for teaming with Wendy and me in making this class a success. This class wouldn't be the same without all of your contributions, including your innovative ideas, design of the website, and help with publication of our journal and scientific posters. I also want to thank Susanne Petri and Nikki Dobos for graciously sharing their classroom/laboratory space with these wonderful students and the many thoughtful conversations they had with the students and me throughout the year. This course and the students were positively impacted by several other RCHS teachers who volunteered their time to help students perform lab protocols and analyze their data: thank you David Ferguson, Matthew Gracey, and Jason Dunkle. I am grateful to the RCHS Department of Science and the school administration for supporting this course, as well as the Douglas County School District for providing the Innovative Education Grant that allowed for the purchase of research-grade science equipment the students used in their research.

I would finally like to thank all of the families, friends, and other donors who contributed to the students' projects. Many of you have asked to remain anonymous, so we will not recognize you by name, but please know how much we appreciate your contributions. For the rest, the students will thank you personally in their individual acknowledgments, but I want to recognize the following donors who donated $100 or more to the students' research this year:

Ed Chan
Kristin Franklin
Sujatha Narayan
Sudha Sharma
Kini Spence
Sandra Tavolario
Lisa Voss
F. G. Watervoort

CONTENTS
April 2017

FOREWORD

This year has been another incredibly successful year of student-led research. I am blown away by the level of scientific research that these students are capable of performing each and every year. They were an amazing group of resilient, hardworking, and thoughtful students, and I feel blessed to have had the opportunity to work with them. The students encountered and overcame many hurdles throughout the year, from learning how to work hard to meet deadlines, how to emotionally handle putting time and so much effort into something only to be told it isn't done or isn't good enough yet, how to problem solve on their own when there was no one who could tell them the next step, and how to work effectively on a team even in stressful situations. Many were surprised at the level of work that was required of them during this initial stage and all were surprised at the level of writing that we expected them to achieve, both in their research proposals and in their final journal articles. It is not easy to believe that these were written by high school seniors, and in the case of one team, high school juniors. We expected them to write and communicate about their research at a graduate level; we believed in their ability to do so, and they rose to this expectation.

As you read their journal articles in this publication you will see what I mean. They have grown as scientists, as writers, and personally in ways that would have been impossible in any other course. At the end of each of their articles, they wrote a reflection about what this experience has meant to them personally and it is these words that make me so proud to have been on this journey with them, and what drives me to continue facilitating this course (I say facilitating because teaching would be the wrong word here).

The big change that was made this year was the addition of a common English course that all research students were enrolled in together. This course, Technical Writing, was led by our wonderful Wendy Lerolland, who was incredibly brave to take on the challenge of supporting students as they learned to read, write, and present like scientists. She supported them throughout the year and together we developed an amazing partnership from which the students and I benefited greatly.

Shawndra L. Fordham
Biotechnology Teacher
Rock Canyon High School
April 11, 2017

When Shawndra Fordham first approached me about teaching an English class to complement the Biotech program, I thought: what fun! Then I looked at the journal, and I thought: No way. No way can I comprehend this stuff. Way over my head; high school biology was 30 years ago: Not Fun.

It was a steep learning curve. I spent more time looking up scientific nomenclature and terminology than grading. I tried to teach what I was learning alongside these students, and it was humbling. I told Shawndra that I didn't understand the content, and she told me that she didn't either, because the students didn't understand the concepts. So we went about learning together, and researching, writing, rewriting, talking through problems, redesigning, rewriting. Shawndra told me that it would get crazy, but it was so much more than that. Despite the tight deadlines, the inevitable revising, the rigor – it was fun.

It's a mystery why these Biotech students would take on this demanding, rigorous, beyond-all-expectations research/ presenting/writing project at this point in high school. It would have been easier to coast through senior year, or postpone the class, in the case of our two junior researchers. But every one of them approached the deadlines with a willingness to put their best effort forward always. They read research journals; they wrote their proposals. They edited their proposals again, and again, and again. They presented their proposals before school board members, parents, principals, peers, and visiting students. Some of their projects were rejected, and some required revision. So they did what was required to continue down this crazy, wonderful, amazing path. It required more than work ethic. As Shawndra often told them, "You can't just be smart. Everybody here is smart." Resiliency? Check. Cooperation? Check. Emotional maturity? Oh, yeah.

Shawndra will say that it's not her doing, it's her students, but I don't know many educators so entirely, 100% devoted, so completely visionary in her approach. As her collaborator and at times co-conspirator, she made sure that my efforts were respected and valued. In my 15th year of teaching, I learned something new: I learned science writing. How cool is that? The program is absolutely crazy and absolutely wonderful, and I am blessed to be along on the ride with these fantastic students and Shawndra's vision. But this wouldn't be complete without a HUGE shout-out to Bryan Winkelman for his technical expertise, grading and maintaining the blogs, assembling the final journal publication, and unfailing patience.

Wendy Lerolland
Technical Writing Teacher
Rock Canyon High School
April 18, 2017

The effects of the IC87114 inhibitor on oligodendrocyte migration in *Danio rerio* embryos

T. M. Rowe, A. M. Chan, A. F. Brown, and S. L. Fordham
Department of Science, Principles of Experimental Design in Biotechnology, Rock Canyon High School, Highlands Ranch, Colorado, USA

IC87114 is a drug that inhibits the p110δ protein, which is one of the many proteins that play a significant role in the phosphoinositide 3-kinase (PI3K) pathway. This pathway is involved in cell migration, growth, and proliferation.[5] Therefore, lowering the expression of the p110δ protein using the IC87114 inhibitor has been found to inhibit the PI3K pathway and its function. Since the PI3K pathway is essential for cell development and migration, we hypothesized that the p110δ protein would be expressed in oligodendrocytes; therefore, exposing *Danio rerio* embryos to the IC87114 inhibitor would lower oligodendrocyte migration to the neural tube during embryonic development. If oligodendrocyte migration is affected, the inhibitor could potentially have off-target neural and developmental effects when used in other treatments. To test our hypothesis, we exposed the *D. rerio* embryos to the IC87114 inhibitor and measured oligodendrocyte migration to the neural tube. We found that the IC87114 inhibitor does have an effect on oligodendrocyte migration. However, contrary to our hypothesis, we found an increase in migration of oligodendrocytes to the neural tube relative to the DMSO control. This indicates that the inhibitor will not have a negative off-target effect on oligodendrocyte migration when used to treat demyelinating diseases. The reason for the increase is unclear, and future research will need to be performed in order to determine the mechanism for this difference. Our research is highly relevant because it did not show evidence to support potential off-target effects on oligodendrocyte migration with the use of the IC87114 inhibitor.

The ability to send messages throughout the body with the help of neurons is essential to the functioning of the human body. The central nervous system (CNS) is a complex system that incorporates many different types of cells, including glial cells, such as oligodendrocytes. Oligodendrocytes are the myelin-producing neuroglia cells found in the CNS.[10] Myelin is a modified plasma membrane that wraps around the axons of nerve cells to insulate them, which speeds up neural impulses and also keeps the impulse from breaking out of the axon **(Pic. 1)**. With proper myelination, signals travel rapidly and the neurons fire off more action potentials.[9] Decreased amounts of oligodendrocytes lower myelination, resulting in slower transmission of nerve impulses and nerve fibers can be more easily damaged. Multiple sclerosis is an example of a demyelinating disease in which the white blood cells (B lymphocytes) attack the myelin sheaths surrounding the axons of neurons as well as oligodendrocytes in the CNS, preventing proper myelination of the neurons and slowing the transmission of messages.[4,9] Neurotransmission is essential to everyday function; therefore, abnormalities in oligodendrocytes and myelination is highly relevant.

Our experiment tests the effect of the IC87114 inhibitor on oligodendrocyte migration **(Pic. 2)**. The IC87114 inhibitor, which is currently being used in cancer research, is similar to the drug Idelalisib, which was approved by the FDA in July of 2014 for treatment of leukemia and two types of lymphoma. These drugs are both p110δ protein inhibitors. The PIK3CD gene, which encodes the p110δ

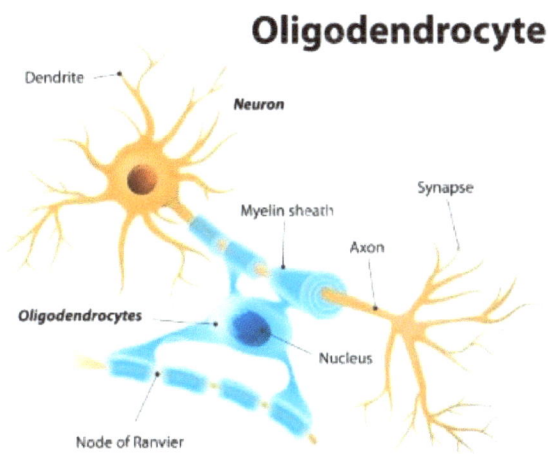

Picture 1: A neuron consists of a cell body (soma), dendrites that receive neural impulses, an axon wrapped in a myelin sheath created by oligodendrocytes, nodes of Ranvier in between myelin, and the axon terminal with a synapse where signals are transmitted to the next axon.[1]

protein, is highly expressed in B lymphocytes, which could result in an excess amount of abnormal white blood cells in leukemia. The inhibitor could decrease the rate of proliferation of white blood cells by inhibiting the p110δ protein and shutting down the PI3K pathway. Idelalisib helps leukemia patients by decreasing the amount of abnormal white blood cells produced.[5] IC87114 is a cell-permeable inhibitor that targets the p110δ protein. When the protein's function is inhibited, it can no longer synthesize signals to allow the cell to migrate.[2]

Picture 2: Chan is making the 7.5uM and 10uM stock solutions of the IC87114 in a fume hood.

The p110δ protein is one of the many proteins involved in the PI3K pathway, which regulates processes involving cell growth, migration, survival, proliferation and apoptosis. The PIK3CD gene plays an important role in the regulation of signaling networks involved with cancer.[7] Research has shown that increased levels of the p110δ protein can result in increased cell migration.[6]

The use of the p110δ inhibitors may be beneficial in treating demyelinating diseases like multiple sclerosis by inhibiting the proliferation of demyelinating immune cells; however, little has been researched on the off-target effects these drugs may have on oligodendrocyte cell migration. We used the IC87114 inhibitor to better understand how the p110δ protein may affect oligodendrocyte migration in *D. rerio* embryos. We hypothesized that if the p110δ protein is expressed in oligodendrocytes, it would be inhibited by IC87114 and would slow the migration of the oligodendrocyte cells to the neural tube in the *D. rerio* embryos. The *D. rerio* embryo lines we used had green fluorescent protein (GFP) tagged oligodendrocytes, allowing us to easily observe their migration to the neural tube.

METHODS
Experimental Design
In order to perform our experiment, we tested 7.5uM and 10uM concentrations of the IC87114 inhibitor, purchased from Echelon Incorporated, on the Olig2:EGFP line of *D. rerio* embryos. These IC87114 concentrations were determined during pre-trials and were tested to have the lowest mortality rates. Embryos were provided by our mentors Tanya Brown and Veronica Fregoso, graduate students in the Cell Biology, Stem Cells and Development graduate program at the University of Colorado Anschutz Medical Campus in Aurora, CO. Our mentors also provided us with embryo media, live imaging media, and tricaine.

Fertilized embryos were selected at four hours post fertilization (hpf) **(Pic. 3)**. At 24 hpf, we sorted the embryos for GFP fluorescence **(Pic. 4)**. We then dechorionated the embryos using fine-tipped forceps to gently puncture and remove the chorion **(Pic. 5)**. After dechorionation, the embryos were transferred to a six-welled petri plate with a CellBIND surface, from Corning Inc., containing 4 mL of embryo media. The six-welled petri plate was maintained within a temperature range of 26°C to 31°C using a terrarium substrate heater from Exo-Terra. Extra layers of insulation were added between the heating mat and petri plate in order to keep the embryos from overheating. The heating mat and embryos were housed inside a laminar flow hood by a window to control their environment with natural lighting and limited disturbance.

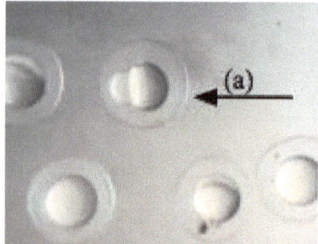

Picture 3: This picture shows the fertilized *D. rerio* eggs under the microscope **(a)**. They are fertilized due to the high oblong cluster of cells above the egg in the chorion.

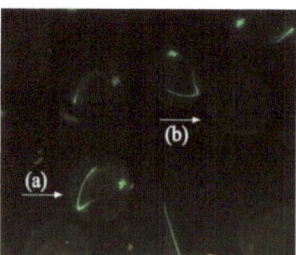

Picture 4: This picture shows the embryos under the microscope with the Nightsea attachment and fluorescent light. The embryos that are glowing green **(a)** have the GFP protein, whereas the ones that are not glowing **(b)** do not.

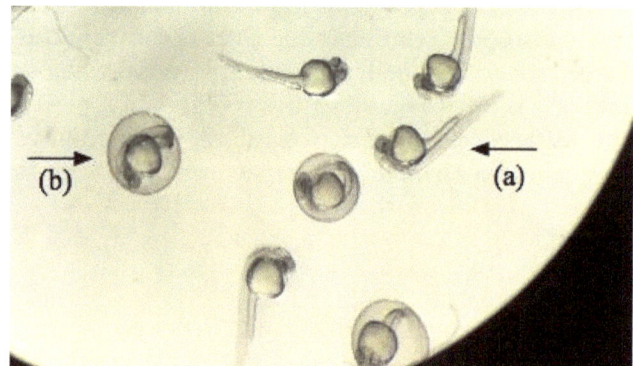

Picture 5: This picture shows both dechorionated embryos **(a)** and embryos without their chorion removed **(b)**. These embryos are 24 hours post fertilization.

To allow the inhibitor to proliferate through the skin of the dechorionated *D. rerio* embryos, we made 750uM and 1000uM dilutions of the IC87114 inhibitor with 1% DMSO and added 40μL of each to the embryo media for a total volume of 4 mL in each treatment. Each well contained a total of 1% DMSO, which was enough to allow the inhibitor to proliferate into the embryos without killing them.

We performed four trials, testing the two different concentrations of the IC87114 inhibitor and the controls. Our first control group contained only the embryo media, while the second control consisted of 1% DMSO and embryo media. The first two treatments consisted of the 7.5uM concentration of the IC87114 inhibitor along with embryo media (C1), and the other two treatments consisted of the 10uM concentration of the IC87114 inhibitor with embryo media (C2) **(Fig. 1)**. We added the IC87114 inhibitor to each treatment at 48 hpf. Each treatment contained approximately 25 fertilized *D. rerio* embryos and 4 mL of the desired treatment.

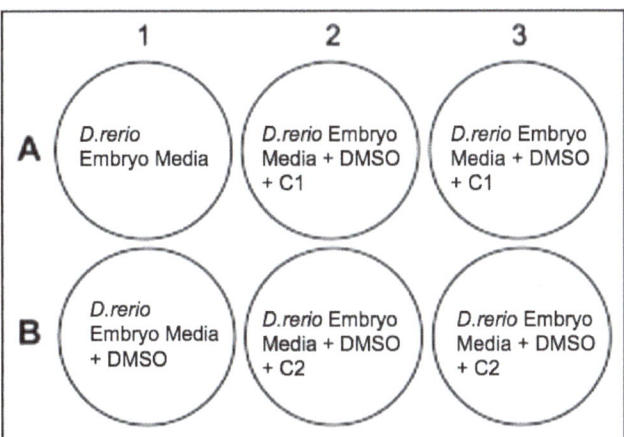

Figure 1: We used three six-well plates pictured in this model for our research. The top left well was the embryo media control and the bottom left well was the DMSO control. C_1 represents the 7.5uM concentration of IC87114 we tested on the embryos and C_2 represents the 10uM concentration.

Imaging

At 72 hpf, we mounted the embryos from each treatment, using tricaine to anesthetize the embryos and live imaging mounting media obtained from our mentors to position the *D. rerio* embryos on their sides. This gave us a clear image of the embryos and their oligodendrocytes under an 80x dissecting microscope **(Pic. 6)**.

Picture 6: In this picture, we are all working on mounting the embryos to take images. Brown (left) is positioning the embryos onto their sides on the microscope slides, Rowe (center) is adding tricaine to the wells, and Chan (right) is adding live image mounting media to the slides.

The oligodendrocytes were tagged with a green fluorescent protein, which allowed us to use a Nightsea fluorescence microscope adapter to visualize and image the oligodendrocytes in the neural tube of each embryo. We used the images taken during mounting to count and document how many oligodendrocytes had migrated to the neural tube at 72 hpf, specifically in the dorsal area across from the yolk sac extension **(Pic. 7)**. At the end of the experiment, embryos were euthanized using ice and placed into biological waste.

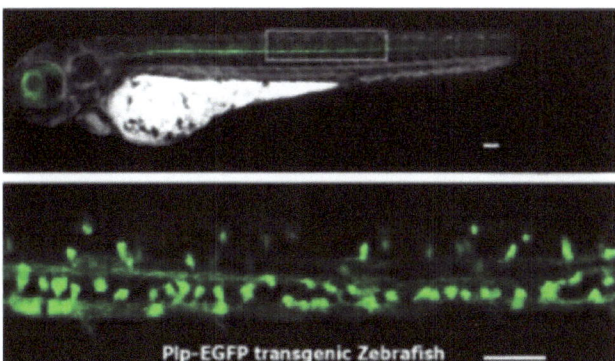

Picture 7: Dorsal oligodendrocyte migration in *D. rerio* embryos. Each small green dot is an oligodendrocyte that has migrated and the box in the top picture is the dorsal area across from the neural tube. This is the area where we counted the oligodendrocytes. (Retrieved from our mentors, Tanya Brown and Veronica Fregoso, used by permission.)

RESULTS

Throughout our experiment, we tested the effects of the IC87114 inhibitor on oligodendrocyte migration to the neural tube in *D. rerio* embryos. We had two control groups, embryo media and DMSO, as well as two IC87114 inhibitor treatments, 7.5uM and 10uM. After counting oligodendrocyte migration, we analyzed, graphed and performed statistical analysis on our data **(Pic. 8)**. We counted 1,127 embryos throughout our entire research, with around 282 embryos per trial. Each researcher counted the oligodendrocyte migration for each *D. rerio* embryo separately and the number was averaged.

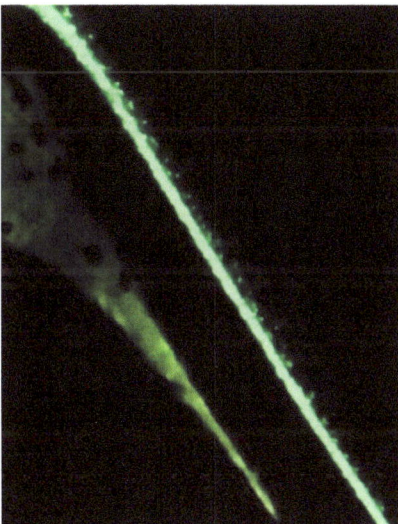

Trial 1 was removed from our data analysis due to low image quality. The GFP protein tagged on the oligodendrocytes was not highly expressed and led to a faint green fluorescence. For these reasons, it was difficult to obtain accurate oligodendrocyte migration counts. Our

Picture 8: This image shows an example of the pictures we took of embryos in each treatment to count the migrated oligodendrocytes. This embryo was imaged from the 10uM concentration of the IC87114 inhibitor + DMSO treatment.

results are expressed as average oligodendrocyte migration for each treatment +/-

1 standard error **(Graph 1)**. Statistical analysis was performed using ANOVA to compare the data among groups, and a T-test was used to compare the treatments and the difference between them. A p-value value of ≤ 0.05 was considered statistically significant.

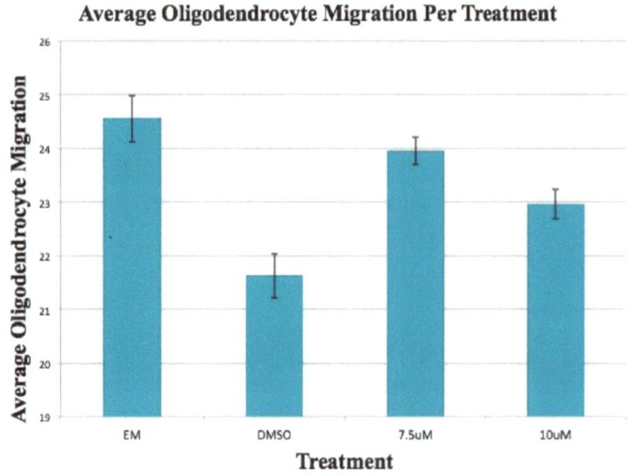

Graph 1: This graph shows the combined average number of migrated oligodendrocytes to the neural tube in *D. rerio* embryos for each treatment in trials 2, 3, and 4. Each treatment from our experiment is shown (bars left to right): embryo media control (EM), DMSO control (DMSO), 7.5uM IC87114 concentration +DMSO (7.5uM) and 10uM IC87114 concentration +DMSO (10uM).

As can be seen in **Graph 1**, the embryo media (EM) control resulted in an average oligodendrocyte migration of 24.6 +/- 0.441 standard error. The DMSO control resulted in an average oligodendrocyte migration of 21.6 +/- 0.414 standard error, which was 3.00 oligodendrocytes lower than that of the embryo media control. This was statistically significant with a significance value of 0.000. Because of this difference, we can see that the DMSO had an overall negative impact on the oligodendrocyte migration to the neural tube in the *D. rerio* embryos. The 7.5uM and 10uM treatments had a lower average oligodendrocyte migration than the embryos observed in the embryo media control group, which could be due solely to the 1% DMSO that allowed the inhibitor to proliferate into the embryos. Therefore, the 7.5uM and 10uM treatments containing 1% DMSO are best compared to the DMSO control and have a statistically significant higher average oligodendrocyte migration in comparison to the DMSO control. The 7.5uM treatment had an average oligodendrocyte migration of 24.0 +/- 0.259 standard error, which was 2.40 oligodendrocytes higher than that of the DMSO control with a significance value of 0.000. In contrast, the 10uM treatment had an average oligodendrocyte migration of 23.0 +/- 0.275 standard error, which was 1.40 oligodendrocytes higher than the DMSO control with a significance value of 0.005.

When the two treatments were compared to each other, the average oligodendrocyte migration of the 7.5uM had a statistically significant increase compared to the 10uM treatment. The 7.5uM treatment was 1.00 oligodendrocyte higher than the 10uM treatment with a significance value of 0.008.

DISCUSSION

Currently, a drug on the market similar to the IC87114 inhibitor is used for the treatment of leukemia, and the IC87114 inhibitor is being researched as a potential treatment for autoimmune diseases. IC87114 stops B lymphocytes from attacking the myelin sheath around neurons by inhibiting the p110δ protein responsible for the regulation of cell growth, development, migration, proliferation, and apoptosis through its role in the PI3K pathway.[5] The inhibitor could also help autoimmune diseases by increasing the oligodendrocyte count, which may help restore the myelin on the neurons. Our research tested the effects of the IC87114 inhibitor in *D. rerio* embryos to determine if the use of this inhibitor could have potential negative off-target effects on oligodendrocyte migration to the neural tube when being used in treatments for demyelinating diseases. If so, the inhibitor could affect the oligodendrocyte migration and myelination of neurons and hinder instead of help with the treatment. We hypothesized that by using the IC87114 inhibitor, which is known to lower the expression of the p110δ protein and PI3K pathway, oligodendrocyte migration would be hindered and we would see decreased numbers of oligodendrocytes that migrated to the neural tube of *D. rerio* embryos. The data we collected could not support this hypothesis.

The average oligodendrocyte migration to the neural tube in *D. rerio* embryos in the embryo media control was 24.6 oligodendrocytes. When exposing the IC87114 inhibitor to the *D. rerio* embryos, we used DMSO in order to increase the absorption. We had a 1% DMSO control to ensure any changes in oligodendrocyte migration were due to the inhibitor and not from the DMSO. For the 1% DMSO control, the average oligodendrocyte migration was 21.6, which was 3.00 lower than that of the embryo media control. This statistically significant decrease observed between the DMSO control and the embryo media control suggests that DMSO negatively affects oligodendrocyte migration in *D. rerio* embryos. When DMSO was combined with the IC87114 inhibitor, the oligodendrocyte migration increased, contradictory to our hypothesis. The average oligodendrocyte migration for the 7.5uM IC87114 inhibitor treatment was 2.40 higher than the average migration of the DMSO control, and the average migration for the 10uM treatment was 1.40 higher than that observed in the DMSO control. Due to the fact that the average oligodendrocyte migration values for the 7.5uM and 10uM IC87114 inhibitor treatments both had a statistically significant increase in average migration when compared to the DMSO control, the DMSO could not have been the only factor affecting the oligodendrocyte migration. We cannot determine why the exposure to the IC87114 inhibitor increased oligodendrocyte migration rather than decreased it.

Previous research provides evidence that DMSO causes apoptotic neurodegeneration in the developing mouse central nervous system.[8] The DMSO could have initiated apoptosis in the developing central nervous system of the *D. rerio* embryos, which may have been the reason for the decrease in average oligodendrocyte migration observed. Previous studies have also shown that DMSO has an influence on activator protein 1 (AP-1) activity.[12] AP-1 is a transcription factor that regulates cell proliferation, transformation, and apoptosis.[11] DMSO may have stopped the cell cycle at the G1 stage, which could explain the lowered amount of oligodendrocytes in the DMSO control.[12]

Explaining the reason that the IC87114 inhibitor increased oligodendrocyte count compared to the DMSO control group is difficult, but one explanation could be related to the GSK3β protein. The PI3K pathway plays a role in regulating the GSK3β protein, which has been found to decrease the differentiation of oligodendrocytes. The GSK3β protein negatively affects oligodendrocyte differentiation, so when inhibited there would be an expected increase in oligodendrocyte differentiation. Inhibiting the PI3K pathway leads to a decrease in the expression of the GSK3β protein, which in turn could lead to an increase of oligodendrocyte differentiation, and migration to the neural tube.[3]

Because the 7.5uM and 10uM treatments showed an increase in oligodendrocyte migration relative to the DMSO control but an overall slight decrease relative to the embryo media control, another explanation is that the IC87114 inhibitor could be protective against the negative effects of DMSO in some way. It is beyond the scope of our research to explain the exact interactions between DMSO, the IC87114 inhibitor, and oligodendrocyte migration, but further studies on the mechanisms behind this at the cellular level would be important to fully understand the results. Without performing more research on how the oligodendrocytes were specifically affected, we are unable to determine what specific process of the PI3K pathway was interrupted by the inhibitor or if it was having an impact through a different mechanism.

Based on our research, the IC87114 inhibitor does not seem to have off-target effects on oligodendrocyte migration and may instead have a positive effect.[9] In order to better understand the mechanism behind our results, further research needs to be conducted. Future studies should focus on oligodendrocyte mitosis and apoptosis. The p110δ protein inhibitor may affect the division of glial cells as well as the ability of the cell to perform apoptosis. It may also be helpful to study the effect of the IC87114 inhibitor on other neuronal cells such as motor neurons and neuroglia stem cells.

Several factors in this experiment may have affected our data. We decided to drop our first trial from consideration because the fluorescence in the oligodendrocytes were not highly expressed, which made it very difficult to adequately count the oligodendrocytes. Our image quality also could have affected our overall results. It was a challenge across all four trials to visualize each individual oligodendrocyte because of the magnification limitations of our Leica KL300 LED microscope. Some of the oligodendrocytes were clustered together and there were other shadows in the pictures, which made it difficult at times to determine the number of individual oligodendrocytes. Lastly, heat inconsistencies caused by the heating mat that the *D. rerio* embryos were kept on, as well as random environmental factors may have affected the oligodendrocyte migration in the *D. rerio* embryos.

ACKNOWLEDGMENTS

First, we would especially like to thank our mentors, Tanya Brown and Veronica Fregoso, graduate students in the Cell Biology, Stem Cells and Development program at the University of Colorado Anschutz, for guiding and supporting us throughout all of our research as well as providing us with *D. rerio* embryos and embryo media. A special thank you to Dr. Wendy Macklin, professor and chair of the Cell and Developmental Biology program at the University of Colorado Anschutz Medical Campus, for donating live imaging mounting supplies and lab reagents to our research. We thank and greatly appreciate all those who donated and supported our research and made it all happen: Alison Cagaanan, Sonia Maaliki, Jennifer Rowe, Linda Humphreys, Laurel Vida, Allan PW Hewett, Maya Harris, Jean Rohrbach, Melissa Kulwicki, Linda Brown, and Steve Wycoff.

We would also like to thank all of the people who helped to guide, support, and contribute to our project by providing us with advice, lab space, and assistance. First, we would like to thank Bryan Winkelman for setting up and designing our research website, guiding us through our blog posts, and providing support to benefit our research. We would like to thank Amy England and Jim McClurg for filming and editing our video for our website. Also, a very big thank you to Wendy Lerolland for her editorial assistance and support throughout our research. We express our gratitude to David Ferguson for helping us with our dilution calculations and providing us with lab equipment to make our solutions, Dr. Jason Dunkle for helping us determine our statistical test and set up our data analysis, Matthew Gracey for assistance with data analysis, and Tom Dillon for providing feedback on our experimental design. We would like to thank Susan Petri and Nikki Dobos for sharing their lab space. Lastly, we thank Rock Canyon High School and the Douglas County School District for providing us with laboratory space and equipment to perform our research.

REFERENCES

1. Admin (2015). Acorda announces phase 1 results for remyelinating antibody in multiple sclerosis. Retrieved 2017, April 18. [Web]
2. Ann Arbor: Cayman Chemical. IC87114 - CAS No 371242-69-2, Cayman Chemical. Retrieved 2016, September 25. [Web]
3. Azim, K., & Butt, A. M. (2011). GSK3β negatively regulates oligodendrocyte differentiation and myelination in vivo. *Glia, 59*(4), 540-553. DOI:10.1002/glia.21122
4. Ben-Nun, A., Mendel, I., Bakimer, R., Fridkis-Hareli, M., Teitelbaum, D., Arnon, R., . . . Rosbo, N. (1996). The autoimmune reactivity to myelin oligodendrocyte glycoprotein (MOG) in multiple sclerosis is potentially pathogenic: Effect of copolymer 1 on MOG-induced disease. *Journal of Neurology*, 243(S1). DOI:10.1007/bf00873697
5. Bouscary, D., Bardet, V., Sujobert, P., Cornillet-Lefebvre, P., Hayflick, J. S., Prie, N. ... Lacombe, C. (2004). Blockade of p110delta isoform activity of phosphoinositide 3-kinase inhibits blast cell proliferation in acute myeloblastic leukemia. *Blood, 104*(11), 2522. DOI:10.1182/blood-2004-08-3225
6. Cain, R. J., & Ridley, A. J. (2009). Phosphoinositide 3-kinases in cell migration. *Biology of the Cell, 101*(1), 13-29. DOI:10.1042/bc20080079
7. Cho, D. (2015). Exploring the pathway: despite lukewarm clinical benefit of PI3K inhibitors, optimism remains regarding biomarkers

and combination therapy. ASCO Daily News. Retrieved 2016, September 25. [Web]

8. Hanslick, J. L., Lau, K., Noguchi, K. K., Olney, J. W., Zorumski, C. F., Mennerick, S., & Farber, N. B. (2009). Dimethyl sulfoxide (DMSO) produces widespread apoptosis in the developing central nervous system. *Neurobiology of Disease, 34*(1), 1-10. DOI:10.1016/j.nbd.2008.11.006

9. Kettenmann, H., & Verkhratsky, A. (2011). Neuroglia, der lebende Nervenkitt. *Fortschr Neurol Psychiatr, 79*(10): 588-597. DOI: 10.1055/s-0031-1281704

10. Robinson, S., & Miller, R. H. (1999). Contact with central nervous system myelin inhibits oligodendrocyte progenitor maturation. *Developmental Biology, 216*(1), 359-368. DOI:10.1006/dbio .1999.9466

11. Shaulian, E., & Karin, M. (2002). AP-1 as a regulator of cell life and death. *Nature Cell Biology, 4*(5). DOI:10.1038/ncb0502-e131

12. Wang, C., Lin, S., Lai, Y., Liu, Y., Hsu, Y., & Chen, J. J. (2012). Dimethyl sulfoxide promotes the multiple functions of the tumor suppressor HLJ1 through activator protein-1 activation in NSCLC cells. *PLoS ONE, 7*(4). DOI:10.1371/journal.pone.0033772

ABOUT THE AUTHORS

Pictured: From left to right, Alexis Chan, Alexi Brown, Tessa Rowe, and our mentor Veronica Fregoso from the University of Colorado Anschutz Medical Campus.
Not Pictured: Tanya Brown, our mentor from the University of Colorado Anschutz Medical Campus.

Over this past year, we have all grown and learned so much individually and as a team. We have learned how to persevere, think critically, and successfully design and perform research. More specifically, we learned how to care for *D. rerio* embryos, dechorionate them, create stock solutions, run statistical analysis, and so much more! Although we ran into a lot of roadblocks, especially toward the beginning of the year in forming ideas for our project, we have all gained a valuable experience by taking this class. When we started, we had a completely different idea for the basis of our project; however, it became apparent that we were trying to incorporate too many ideas together that didn't connect. Although we wanted to relate schizophrenia, oligodendrocyte migration, and the IC87114 inhibitor together there was no way to do so. We spent lots of time digging through literature and trying to comprehend the concepts, until we finally decided on our research idea. The formation of our project caused a great amount of frustration, but it taught us perseverance, hard work, and determination to keep trying to understand the complex material. We were put into real world situations and face challenges that real scientists encounter. This knowledge will help us further our education at the college level and open doors for future job opportunities. Unlike other high school students, we were given the opportunity to go above the average high school curriculum and extend our knowledge beyond a classroom or textbook.

We are excited to be taking this experience with us to the college, or university, we have chosen. This class has allowed us to explore different career paths in biotechnology. Beyond our research, we have learned how to communicate with each other and professionals in the biotechnology field, including our mentors, whom we saw almost weekly for assistance. We are so grateful for this amazing opportunity to learn valuable life skills and experience individual lab work. We had lots of fun this year and this program will always have a place in our hearts!

Effects of *Hericium erinaceus* on induced neurodegeneration in *Danio rerio* embryos

N. Chauhan, E. N. Hadjis, and S. L. Fordham
Department of Science, Principles of Experimental Design in Biotechnology, Rock Canyon High
School, Highlands Ranch, Colorado, USA

***Hericium erinaceus*, the lion's mane mushroom, is used in ancient Chinese medicine and many recent studies have shown that it has neural regenerative properties. In this experiment, we tested the effects of an aqueous extract of *H. erinaceus* on neurological development in *Danio rerio*, zebrafish, embryos. We measured the sensory responses and general morphology of the embryos after they had been exposed to a neurodegenerative chemical and the mushroom extract. This test was designed to show whether an extract from the mushroom could prevent damage to the embryos' neurological systems from a degenerative chemical: in this case, ethanol. Since *H. erinaceus* has been observed in previous research to stimulate peripheral nerves' regrowth after they have been physically damaged, we hypothesized that the embryos would exhibit less neurological damage when exposed to both the harmful chemical and the mushroom extract than when exposed to only the ethanol. We measured the embryos' neurological damage using two tests that evaluated their energy/reactivity and ease of movement when touched with a probe. The reactivity test did not result in statistically significant data since the embryos' reactions in the mushroom extract showed an improvement or a higher reactivity score as compared to the ethanol-only control, but this was not a large enough difference to strongly correlate an improvement in health caused by the mushroom. The ease of movement test resulted in statistically significant data that showed an improvement in the embryos movement when exposed to the mushroom extract, as they showed ratings similar to those of the embryos exposed to no chemicals. Overall, these results support our hypothesis that the lion's mane mushroom may be capable of prophylactically protecting against neurological damage.**

Hericium erinaceus originated in East Asia and has a long history of use in Chinese herbal remedies (**Pic. 1**). Known colloquially as the lion's mane mushroom, it has been traditionally used for its reported antibiotic, anti-fatigue, neuroprotective and antidiabetic properties, as well as the ability to improve cognitive function and mitigate anxiety.[1] Studies indicate that these properties may be linked to the mushroom's antioxidative, immunostimulatory, and anti-inflammatory properties.[11] The vast potential health benefits of *H. erinaceus*, as well as its curious lack of observed side effects, make it promising in having positive effects in a regenerative model. Studies of the mushroom's healing properties have shown its potential to prevent against neurodegenerative diseases like Parkinson's disease (PD) and Alzheimer's disease, disorders associated with age.[5]

A study conducted in 2014 on Sprague-Dawley rats showed that *H. erinaceus* contributed to the repair of damaged peripheral neurons.[11] A different study in 2010 indicated that hericenones and erinacines, which are compounds in the mushroom's fruiting bodies and mycelium, respectively, stimulate cell production of nerve growth factor.[4] These results indicate *H. erinaceus*'s potential to aid with nerve regeneration. Likewise, research has shown that improving myelination or speeding up neuron development are alternative methods in relieving neurodegenerative symptoms.[2]

Neurodegenerative diseases have far-reaching effects: PD alone affects over 1 million people in North America.[3] Their symptoms include progressive loss of control over motor functions, resulting in stiff, slow and/or unstable, jerky movement.[11] These effects can be modeled in zebrafish embryos by introducing a neurologically damaging chemical such as ethanol. As ethanol causes lesions (sections of damaged tissue) in the central nervous system, testing whether an extract from *H. erinaceus* can protect *D. rerio* embryos from neurological damage induced by ethanol would indicate whether this mush-room is viable as a prevention for neurodegenerative diseases.[9] A touch assay examining the change in the motor function of the embryos was used in this experiment as the measure of neurological damage.

D. rerio embryos have structural similarities to mammalian neurological systems.[6] We conducted behavioral and developmental assays and morphological observations on the zebrafish embryos as a measure of the phenotypic neurological effects of the mushroom extract. These tests evaluated the mushroom's protective effects, not actual regeneration, on the peripheral and motor neuron systems

Picture 1: *Hericium erinaceus* mushroom growing on a tree, its natural environment.[8]

because the embryos were not allowed a significant recovery time for repair.

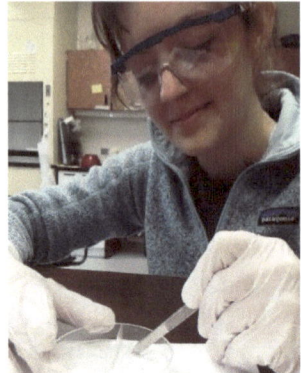

To measure the mushroom's effects, we used scales to rate the reactivity and ease of movement of the embryos' reactions to touch, then compared these measures to those of the embryos in the control trials **(Pic. 2)**. This method has been validated by a related study on *D. rerio* in which researchers used touch assays to classify the neurological function of the embryos by observing how dramatically the embryos responded to a touch by a probe.[7] We also recorded whether the embryos developed into the expected stages at the expected times and noted any major morphological changes in the embryos that were observed.

Picture 2: Hadjis conducting the touch assays on zebrafish embryos.

METHODS
We conducted preliminary trials to determine which of the three chemicals – Roundup, toluene, or ethanol – could best induce clear neurological damage in the zebrafish. We identified that 40 µL of ethanol per well produced the most observable neurological changes in the embryos without killing them.

Picture 3: (a) The dried *Hericium erinaceus* before it was extracted into ethanol. **(b)** Here, Hadjis holds up one of the dried mushrooms, showing its relative size.

Extract Preparations
Before beginning pre-trials, we made an extraction of the lion's mane mushroom obtained from Mathew Ward with the Laboratory Services Division of the Colorado Department of Public Health and Environment **(Pic. 3)**. We ground the mushroom using a food processor to form a fine powder and prepared a 1:10 mushroom-ethanol mixture.

After centrifuging the mixture, only the supernatant was removed and kept for experimentation. We created a 1:1 dilution of this extraction with embryo media.

Experimental Design
We conducted three experimental trials over three weeks, with one trial per week. For each trial, we conducted three control and two experimental treatments **(Pic. 4)**. The first control consisted of 7 mL of 5.3-mM NaCl zebrafish embryo media, while the second control had 7 mL of embryo media plus 40 µL of 99.7% dimethyl sulfoxide (DMSO), and the third control had the same volumes of embryo media and DMSO, as well as 80 µL of 100% ethanol. These controls were used as baseline data for the behavior, development and overall morphology of the *H. erinaceus*-treated embryos. DMSO was added to allow the embryos to absorb and uptake the other chemicals. Each of the two experimental treatments was composed of 7 mL of embryo media, 40 µL of 99.7% DMSO and 80 µL of 100% ethanol, in addition to the mushroom extract or dilution. In the first treatment, we added 40 µL of a 1:10 mushroom extraction, and in the second treatment we added 40 µL of a 1:1 dilution of the extraction.

In each trial, 20 embryos were placed into each of the five wells for a total of 100 embryos per trial **(Pic. 4)**. Temperature was maintained within a range of 26°C to 31°C using a Sunbeam 722-810 electric heating pad with added layers of insulation between the heating pad and the six-welled plate.

The embryos atop the heating pad were housed inside a laminar flow hood to create a controlled environment and prevent them from being disturbed. They also were kept in a classroom with windows to ensure exposure to natural light cycles.

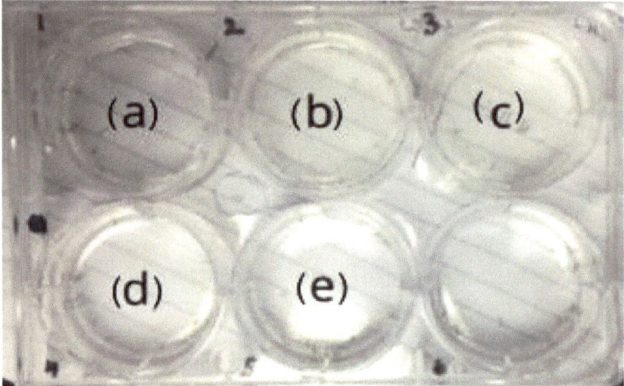

Picture 4: This six-welled plate held 20 embryos per well. Well **(a)** had media only, well **(b)** had media and DMSO, well **(c)** had media, DMSO and ethanol, well **(d)** had media, DMSO, ethanol and the pure mushroom extraction and well **(e)** had media, DMSO, ethanol and the mushroom dilution.

Experiment
We obtained wild type zebrafish embryos from our mentors Anthony Junker and Erik Linklater, graduate students from the Cell, Stem Cell and Developmental Biology Program at at the Anschutz Medical Campus at the University of Colorado, Denver. We collected them 4 hours post fertilization (hpf), and then removed all unfertilized

embryos at the Rock Canyon Biotechnology Laboratory. At approximately 28 hpf, we dechorionated the embryos and separated them into their treatment wells in the six-welled plate **(Pic. 5)**.

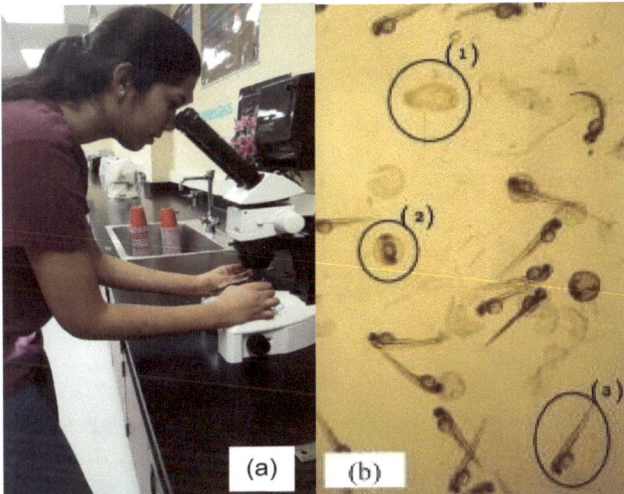

Picture 5: (a) In this picture, Chauhan dechorionates zebrafish embryos. **(b)** Taken through the eyepiece of a dissection microscope, this picture shows **(1)** an empty chorion after dechorionation, **(2)** an embryo prior to dechorionation and **(3)** an embryo after dechorionation.

At 28 hpf we added the mushroom extracts and the DMSO **(Pic. 6)**. At 2 days post fertilization (dpf), we added the ethanol to wells c-e **(Pic. 4)**. At the same time of day for the next two days, we recorded morphological observations. At 4 dpf, we performed touch assays.

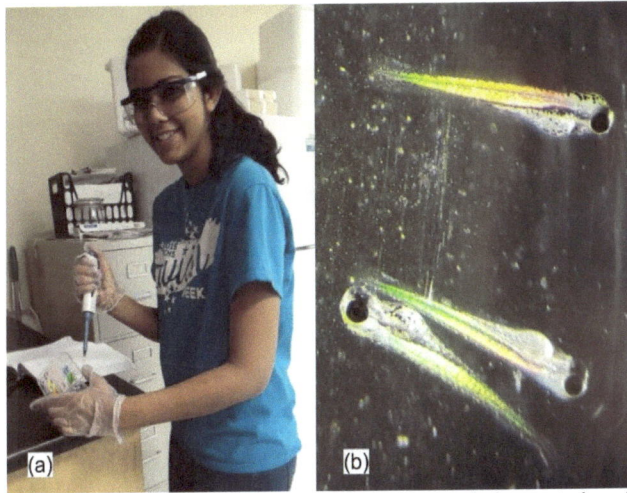

Picture 6: (a) Chauhan adds ethanol to some of the control and experimental wells. **(b)** A view of embryos after dechorionation and prior to exposure to ethanol.

Touch Assays

We rated the energy and ease of movement during each embryo's reaction to a gentle nudge with the blunt end of a forcep on a scale of -5 to 5 **(Pic. 2)**. A 0 on this scale was based on the reactions of embryos observed in our media-only control treatment. A positive score correlated with a more energetic reaction, while a negative score resulted from a less dramatic reaction or less ease of movement.

Less ease of movement was observed as jerkiness, inability to swim in a straight path and sometimes even near-paralysis. Immediately after collecting data on the fourth day, we euthanized the embryos to ensure compliance with Institutional Animal Care and Use Committee (IACUC) regulations. To euthanize them, we placed the embryos on ice in a -20°C freezer.

RESULTS

After collecting data, we first found the average ratings for both the embryos' energy/reactivity and their ease of movement in each trial on a scale from -5 to 5, with 0 indicating a normal reaction. A positive reactivity rating would indicate that an embryo swam in a greater number of bursts, meaning that they were more reactive than a typical embryo with a rating of 0, while a negative reactivity rating would indicate swimming for less time or a smaller distance after a nudge. A higher ease of movement than the normal rating would occur if the embryos had a greater range of motion or swam a greater distance with less effort than normal embryos. A negative ease of movement rating meant embryos wiggled their tails and fins vigorously and did not move as far or as quickly as embryos with a rating of 0. For our three control trials, we exposed zebrafish embryos to only media; to media and DMSO; and to media, DMSO and ethanol. Our two experimental trials both consisted of embryo media, DMSO and ethanol, as well as a pure mushroom extract for our first experimental trial and a dilution of the mushroom extract for our second.

We averaged the touch assay ratings for the reactivity and ease of movement tests from each trial and performed tests on them. We used a 95% confidence interval in order to determine whether experimental treatments showed statistically significant differences from the control treatments.

It is important to note that we found that the embryos in the embryonic media and DMSO control did not have significantly different ratings in either reactivity or ease of movement from the embryos in the baseline media-only control. This shows that any changes between the media-only control and experimental trials were due solely to the mushroom treatment, and not to the effects of the DMSO.

The embryos in the ethanol control had an average reduction in reactivity of 0.9375 as compared to the baseline of zero for embryos exposed to only embryo media control, with a significant difference of 0.0027%. This shows that ethanol was effective in having the desired negative impact on the neurological health of the embryos. However, the embryos' reactivity in each of the experimental mushroom treatments averaged -0.02632 and 0.27027 for the extract and dilution of the extract, respectively **(Graph 1)**. This showed an overall improvement in reaction as compared to the embryos in the ethanol control trial. However, this difference was not statistically significant. The embryos in the ethanol control were very unresponsive to touch; often, they would have a delayed or reaction or a very short reaction, whereas a normal embryo would react more immediately and for a longer period of time when touched with the probe.

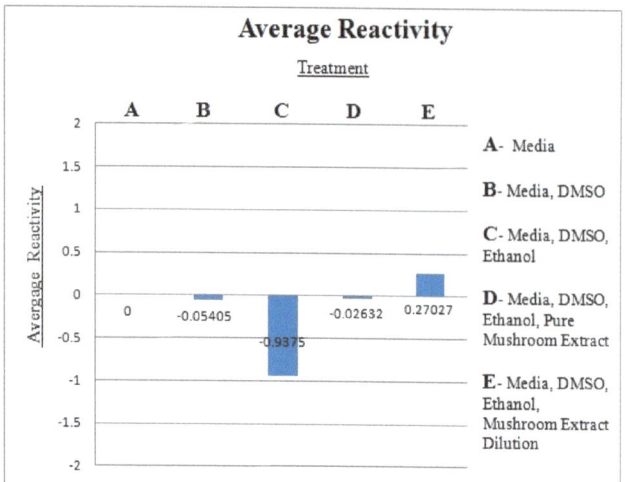

Graph 1: This graph visually and numerically shows the average reactivity or energy ratings of the embryos in each experimental and control trial.

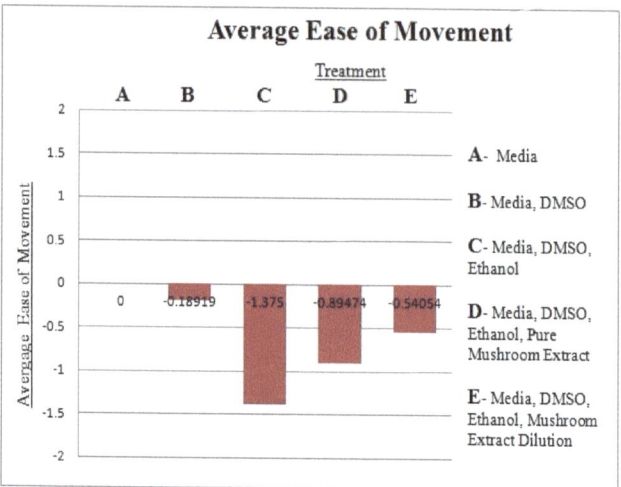

Graph 2: This graph visually and numerically shows the average ease of movement ratings of the embryos in each experimental and control trial.

Our results for the ease of movement ratings had different implications. The embryos exposed to the pure mushroom extract and ethanol had an average reaction to touch rating of -0.805, which is significantly lower than a normal reaction of 0, but is significantly higher than -1.375, which was the average reaction of the embryos in the ethanol control **(Graph 2)**. These embryos exhibited some difficulty swimming but could generally swim in coherent patterns. The embryos that were exposed to the diluted mushroom extract had an even higher average ease of movement of -0.5405. The reaction of the embryos in this well was about the same as the reaction of the ones in the well with the pure mushroom extract: they tended to show difficulty moving but overall could swim without much effort. The reactions of the embryos in the mushroom extract wells were not statistically different from each other, so the concentration of the mushroom extract did not affect the ease of movement of the embryos. The normal cumulative distribution function test showed in this case a percent confidence of 99.94% for the pure mushroom extract and ethanol and 98.46% for the well with mushroom dilution and ethanol treatment. Since this is above 95%, the data is statistically significant. However, even though this data shows that there was a significant improvement in the movement of the embryos, this reaction was still significantly lower than a normal reaction. While the mushroom did help improve the embryos' movement, it did not fully bring it back to the normal reaction of the embryos in the DMSO control.

We also recorded general morphological differences at 3 dpf. In the ethanol control, pure mushroom extraction and mushroom extract dilution experimental treatments, 27 out of 35 total, 13 out of 38, and 7 out of 36 total embryos respectively had enlarged hearts, which was statistically significant and was not observed in the DMSO and media-only controls **(Pic. 7)**. Though the relation of this to the neurological impacts of the ethanol and mushroom is unknown, , it suggests the harmful impact of the ethanol on the embryos' other anatomical systems. It also suggests that the mushroom extract dilution but not the pure extract may

the embryos' other anatomical systems. It also suggests that the mushroom extract dilution but not the pure extract may have potentially helped prevent against these effects because the number of embryos with enlarged hearts was less in the mushroom extract treatment. This reduction was found to be statistically significant because their chi-squared value of 20.829 was greater than the 95%-confidence p-value of 3.84. Several of the embryos in the pure mushroom extract treatment also had a significant number of enlarged hearts, but it was more similar to the ethanol-treated wells and was not statistically different.

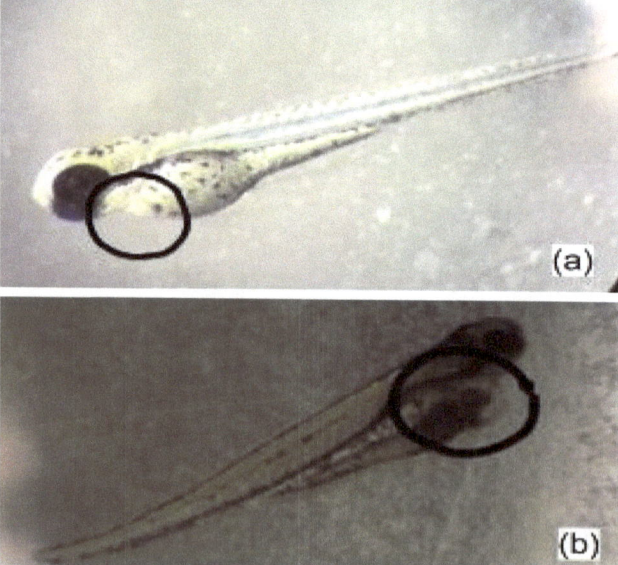

Picture 7: (a) This picture shows a normal embryo with an average heart size. **(b)** When exposed to ethanol, the embryos showed an enlarged heart, as circled in the picture above.

DISCUSSION

In earlier studies, the lion's mane mushroom has been shown to repair damaged neurons, which potentially could help mitigate the symptoms of neurodegenerative diseases.[1] We tested whether the lion's mane mushroom can protect against the damage caused by exposure to ethanol in zebrafish embryos. To test this, we first exposed the embryos to a mushroom extract and a dilution of the extract, then to ethanol, and later examined how the embryos reacted to two different touch assays. The touch assays separately analyzed two aspects of embryos' movement: reactivity/energy and ease of movement. These were used as indicators of the embryos' motor neuron health, and therefore as indicators of their overall neurological health. A higher or more positive rating was correlated to better neurological health.

We hypothesized before experimentation that the mushroom extract and dilution of the extract would protect the embryos' neurological systems from damage caused by ethanol. Such a result would translate into touch assay ratings closer to zero when the embryos were exposed to both the mushroom extract and ethanol, than when they were exposed to ethanol alone. A touch assay rating closer to 0 would show that the embryo reacted more like a normal embryo. Our controls served to show that any results were due to the effects of the mushroom extract or dilution of the extract alone. We found no statistical difference between the reactions of the embryos in the media only control and the DMSO control. This showed that the DMSO, which was also used in the mushroom extract and dilution treatments to increase the amount of ethanol and mushroom extract that would be absorbed by embryos, was not a contributing factor to the observed results in this experiment.

Ease of movement was determined by evaluating how effortless and smooth – or not – the embryos' swimming was when they were touched with the back end of a pair of forceps (Pic. 2). For instance, if an embryo exerted great effort without moving far or could only swim in a circular pattern, then it was given a negative score. We rated embryos negatively when they wiggled their tails vigorously without swimming very far. Others sporadically and uncontrollably changed directions instead of swimming in straight bursts. Our results showed a significant improvement in ease of movement when we exposed the embryos to both concentrations of the mushroom extract. The embryos' average reaction when exposed to ethanol alone was -1.875. Meanwhile, when the embryos were exposed to the mushroom extract in addition to the ethanol, the average reactions were -0.80474 and -0.54054. Since we correlated a higher rating with better neurological health, this result supports our hypothesis and shows potential prophylactic protective qualities of the mushroom.

On the other hand, the ratings for reactivity did not significantly improve between embryos exposed to the ethanol alone, and those exposed to the mushroom extract treatments. Though the results, as seen in **Graph 2**, show that the embryos in both the mushroom extract and extract dilution treatments showed an improvement in reactivity compared to that of the ethanol-only control, it was not enough of a difference to be statistically significant.

Though only one of the tests of neurological health, ease of movement, resulted in statistically significant data, there was a positive improvement on both assays for embryos exposed to mushroom as compared to the embryos exposed

to only ethanol. This data supports our hypothesis. It is important to note that the ease of movement assay was considerably easier to conduct because the ratings seemed to be less subjective and more specific whereas reactivity had too many factors and resulted in discrepancy on what constituted a certain rating.

Throughout this experiment there were some factors that impacted our results. In one trial, there were unexplainable deaths of embryos in our mushroom extract treatments and our DMSO control treatment, which are completely unrelated wells. We think these deaths were probably due to a bacterial contamination in the original media from which we obtained those embryos. When we controlled for this factor, we did not have unexplained deaths. We also believe we saw little impairment of movement in the ethanol control wells for one trial because we used a different bottle of ethanol in this trial than in others. The ethanol in this may have been old and less potent than the ethanol we used for the rest of our trials and therefore caused fewer health effects. Due to this issue we did not use data from this trial in our results.

We suggest that any future experimentation related to this research collect more data to validate our results, and that neurodegeneration is more directly quantified. Such tests might, for instance, quantify the embryos' brain development or more accurately test their movement capabilities by measuring their acceleration while they swim. A similar research study testing toxicity in embryos, for instance, evaluated them using standardized touch assay equipment that delivered stimuli at regular intervals using solenoids.[10] Testing could be extended into the embryos' adult stages of life, with the proper IACUC approval, in order to test the effects of the ethanol and mushroom extract over a longer period of time to more accurately measure potential neurodegeneration. Results from such studies could be used to draw stronger conclusions about the potential capabilities and uses of the lion's mane mushroom in both treating and preventing against neurological harm.

ACKNOWLEDGMENTS
We would like to thank our mentors at the Anschutz Medical Campus of the University of Colorado Denver, Erik Linklater, Anthony Junker and Tanya Brown, for supporting us in designing this project and for generously providing the zebrafish embryos and embryo media we used. We would also like to thank Matthew Ward and the Laboratory Services Division of the Colorado State Department for generously providing the dried *Hericium erinaceus* samples used in this investigation. We would like to thank Susanne Petri and Nikki Dobos, the science teachers who have shared their laboratory space with us. We would also like to thank Rock Canyon High School and Douglas County School District for providing the laboratory and equipment we used for this experiment. We thank all of the other teachers who have lent us their support throughout this project, including David Ferguson for his support with identifying potential neurodegenerative chemicals, Wendy Lerolland for her extensive help with editing this article and Tom Dillon for his feedback and

support with our experimental design. We would like to thank Bryan Winkelman for his support with our blog posts and pitch presentation as well as his help with mentor connections and communication. We would also like to thank Jim McClurg and Amy England for their support in filming our promotional videos.

REFERENCES

1. Friedman, M. (2015). Chemistry, nutrition, and health-promoting properties of *Hericium erinaceus* (lion's mane) mushroom fruiting bodies and mycelia and their bioactive compounds. *Journal of Agricultural and Food Chemistry*, *63*(32), 7108-7123.
2. Frostick, S. P., Yin, Q., & Kemp, G. J. (1998). Schwann cells, neurotrophic factors, and peripheral nerve regeneration. *Microsurgery*, *18*(7), 397-405.
3. Lang, A. E., & Lozano, A. M. (1998). Parkinson's disease. *New England Journal of Medicine*, *339*(15), 1044-1053.
4. Ma, B. J., Shen, J. W., Yu, H. Y., Ruan, Y., Wu, T. T., & Zhao, X. (2010). Hericenones and erinacines: stimulators of nerve growth factor (NGF) biosynthesis in *Hericium erinaceus*. *Mycology*, *1*(2), 92-98.
5. Panula, P., Sallinen, V., Sundvik, M., Kolehmainen, J., Torkko, V., Tiittula, A., ... & Podlasz, P. (2006). Modulatory neurotransmitter systems and behavior: towards zebrafish models of neurodegenerative diseases. *Zebrafish*, 3(2), 235-247.
6. Phan, C. W., David, P., Naidu, M., Wong, K. H., & Sabaratnam, V. (2015). Therapeutic potential of culinary-medicinal mushrooms for the management of neurodegenerative diseases: diversity, metabolite, and mechanism. *Critical Reviews in Biotechnology*, *35*(3), 355-368.
7. Ribera, A. B., & Nüsslein-Volhard, C. (1998). Zebrafish touch-insensitive mutants reveal an essential role for the developmental regulation of sodium current. *The Journal of neuroscience*, *18*(22), 9181-9191.
8. SKas (Own work) [CC BY-SA 4.0 (http://creativecommons.org/licenses/BY-SA/4.0)], via Wikimedia Commons
9. Sledge, D., Yen, J., Morton, T., Dishaw, L., Petro, A., Donerly, S., ... Levin, E. D. (2011). Critical Duration of Exposure for Developmental Chlorpyrifos-Induced Neurobehavioral Toxicity. *Neurotoxicology and Teratology*, *33*(6), 742–751.
10. Thomson, A. D., Ryle, P. R., & Shaw, G. K. (1983). Ethanol, thiamine and brain damage. *Alcohol and Alcoholism*, *18*(1), 27-43.
11. Wong, K. H., Kanagasabapathy, G., Naidu, M., David, P., & Sabaratnam, V. (2014). *Hericium erinaceus* (Bull.: Fr.) Pers., a medicinal mushroom, activates peripheral nerve regeneration. *Chinese Journal of Integrative Medicine*, 1-9.

ABOUT THE AUTHORS

Pictured: From left to right, our mentors, Anthony Junker and Erik Linklater, with Elise Hadjis and Neha Chauhan at the research laboratory on the University of Colorado Anschutz Medical Campus.

I (Neha) am extremely grateful to have had the opportunity to conduct a research project while only in high school. It has been a privilege to gain such in-depth insight into the research process, especially because I plan on studying bioengineering when I begin college in the fall. This project has been an experience I will be to apply to my future endeavors not only because of the scientific principles I've learned, but also because of the perseverance and collaborative skills it has taught me.

I (Elise) am very glad to have been given such an amazing opportunity. Science has always fascinated me and to have the chance to conduct novel research. I have learned a lot about working on a team and problem solving through this. However, I think the biggest thing I learned was how to persevere when things don't go as planned, as they often don't. These skills will greatly help me as I move on to college and in the future to problem solve and work collaboratively in teams.

Using Pluronic F127 to 3D bioprint the hepatic portal vein of the human liver

S. B. Martin, B. C. Timmons, and S. L. Fordham
Department of Science, Principles of Experimental Design in Biotechnology, Rock Canyon High School, Highlands Ranch, CO, USA

Although liver transplants are very commonly needed, livers are not always available. The ability to 3D bioprint a fully functioning liver creates opportunities to ensure that a transplant is more readily available, but this isn't possible without being able to bioprint the vascularization of the liver. 3D bioprinting is an emerging new technology with potential to create functional artificial human organs and tissues. Using a r3bEL 3D Bioprinter, we designed and attempted to print the vascular structure of the hepatic portal vein of the human liver. Printing the vessel horizontally was not successful because we were unable to create supports to form the cylindrical structures. We decided to test the feasibility of a new method for printing vascular structures by printing vertically. We thought that if the vessels could be printed vertically and maintain their cylindrical shape, they could be pieced together to form the entirety of a blood vessel. We used the bioink Pluronic F127 to print these structures. To test this method, we printed cylinders of varying diameters, comparing height to structural integrity. The 6 mm diameter cylinders reached an average height of 14.7 mm, the 7.5 mm diameter cylinders reached an average height of 22.15 mm, the 9 mm diameter cylinders reached an average height of 27.15 mm, and the 10.5 mm diameter reached an average height of 23.85 mm while remaining structurally sound. Our data showed that a wider diameter correlated with a larger vessel until the diameter grew to above 10 mm, where height began decreasing. This unexpected decrease can be attributed to the change in print pattern, but could be changed with a different program. We found that, with some modification, this new method could be a viable option for printing vascular structures in the future.

At any given time, approximately 118,000 people are approved for organ transplants, and 76,000 of those patients are on the waitlist.[12] A new technology, 3D bioprinting, has the potential to produce functional living organs, blood vessels, and tissues for the human body the could address the need for organs. This technology combines cells and biocompatible materials into extremely thin, printable layers that form organs and/or tissues. Bioprinters give the user high-precision control over the placement of biological materials on a flat, 3D grid.[15]

Scientists across the world have been experimenting with 3D bioprinting to fully understand its potential. The Jennifer Lewis Lab at Harvard University has researched bioprinting vascular networks with different bioinks. They found that although 3D bioprinting is a promising method for future use, cell viability remains an issue with current materials.[6] Other labs are experimenting with 3D bioprinting and our understanding of this new technology continues to grow. For example, scientists with several universities are working in collaboration to print a 3D bioengineered placenta while other researchers are working to use 3D printing to regenerate nerve tissue for implantation.[7,10] Researchers at Brigham and Women's Hospital are working to develop artificial blood vessels using hydrogel constructs.[2] We are just beginning to understand the capabilities that 3D bioprinting has to offer, and our research will help to discover more possibilities.

3D bioprinters differ from a regular 3D printer in their

Picture 1: This is Timmons (front) and Martin (back) working with the r3bel 3D Bioprinter as it is printing one of our designs.

ability to print materials that can support life. Normal 3D printers have the ability to make three-dimensional solid objects based on a digital file created by the user. The object is created using additive processes, which means that it is printed layer by layer, with each layer a thinly sliced horizontal cross-section of the final product. The initial model is designed using Computer Aided Design (CAD) 3D modeling software. Next, the 3D model is prepared for printing by a process called slicing, which divides the created model into separate layers. Once the model is sliced, it is ready to be sent to the printer. Although the printer reads every slice as a 2D model, printing layer by layer ultimately leads to a 3D structure.

Picture 2: The r3bEL 3D Bioprinter we used for our research, along with Martin and Timmons.

The main feature of a 3D bioprinter is its ability to print with materials that promote cell growth, known as bioinks. Alginate hydrogels are seaweed-based bioinks made from 99% H_2O, while polycaprolactone (PCL) is a biodegradable thermoplastic commonly used in medical applications and 3D bioprinting.[4,13] While both of these inks would work well for our research, we chose to use Pluronic F127 because it was shown in the Jennifer Lewis Lab to be effective with cell growth of vascularized tissues. It also dissolves in water, which sets it apart from several other inks.[9]

Of the 118,000 people approved for organ transplants, 17,000 are in need of a liver transplant.[11] The human liver is one of the most commonly transplanted organs. The liver breaks down, balances, and creates nutrients for the blood to ingest into the body, while also cleaning the blood of harmful substances. Heavy stress on the organ may lead to the need for a transplant later in life.[5] Hepatitis C and long-term alcohol abuse are among the leading causes of cirrhosis of the liver. Cirrhosis of the liver results in weakness, fatigue, loss of appetite, nausea, vomiting, and weight loss, and without a transplant in severe cases the patient typically only lives up to 90 days.[14] Of the 17,000 people each year in need of a liver transplant, only approximately 6,000 people will receive one and 1,500 will die waiting.[11]

While there is an explosion of research on 3D bioprinting in the scientific field, very little research has specifically been done on 3D bioprinting of the liver. In order to print a fully functional bioprinted liver, though, scientists need to be able to print its vascularization. Creating the vascular structures associated with an organ is a major challenge for scientists in organ engineering. By working to design and print the vasculature of the liver, we will be advancing the science of 3D bioprinting and providing a key component in the future of organ transplants.

In our research, we attempted to print a scaled down version of a portion of the vascularization of the liver, the hepatic portal vein, using the r3bEL 3D Bioprinter by SE3D Education (Pic. 1 & 2). The r3bEL prints using extrusion based technology, thus it has a syringe that needs to be filled and replaced every few prints (Pic. 3).

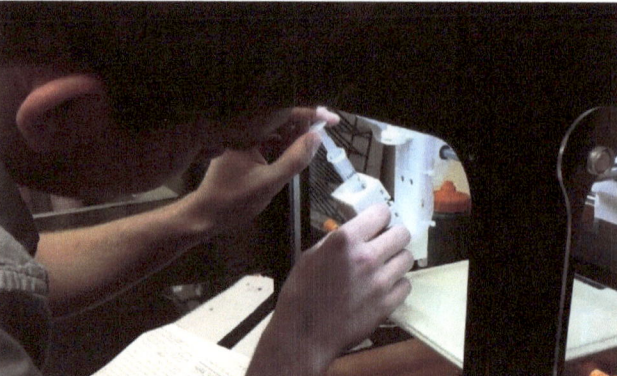

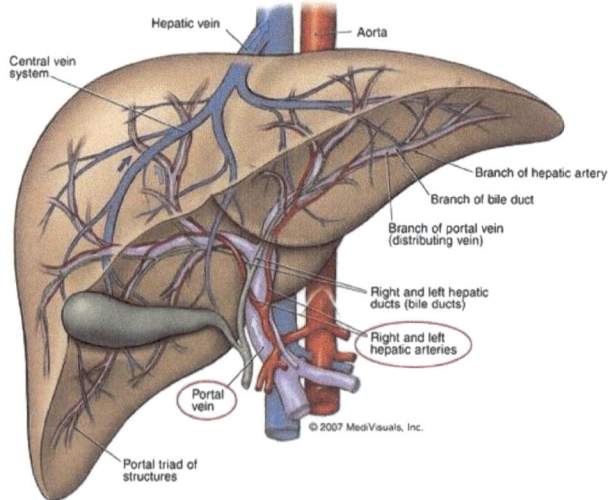

Picture 3: Martin changes the syringe in the extruder on the r3bEL 3D Bioprinter in between prints.

The hepatic portal vein is the biggest blood vessel feeding into the liver and carries the nutrient rich blood from the stomach and intestines to the liver for cleansing (Fig. 1). Our goal was to identify the effectiveness of this specific printer to accurately print this structure. In addition, we tested a new method of printing vascular structures by printing individual portions vertically and piecing them together to make a complete vessel. We tested the height and accuracy of these printed structures to determine if vertical printing is a viable method for future use.

Internal Anatomy of Liver

Figure 1: The anatomy of the liver. We attempted to replicate the portal vein that enters through the bottom, shown in purple.[1]

METHODS/RESULTS

Throughout this experiment, we used the computer applications Arduino, TinkerCad, Slic3r and Pronterface to design and attempt to print a scaled down version of the vascular structure of the hepatic portal vein in the human liver using Pluronic F127 bioink. The hepatic portal vein is

11.545 ± 1.514 mm in diameter at the entry to the liver, and gets smaller as it branches until it reaches approximately 6 mm, which are the sizes we created for our 3D models.[10] We designed the structure at a 1:6 scale. This scale was selected because the r3bEL can print as small as 1 mm thick, and the smallest vessel we were attempting to print was 6 mm in diameter.

Design One

The first design that we created was a simple vessel-like structure with a few branching parts, to test the printer's ability to print a horizontal structure before we tried to print the whole vein (**Fig. 2**).

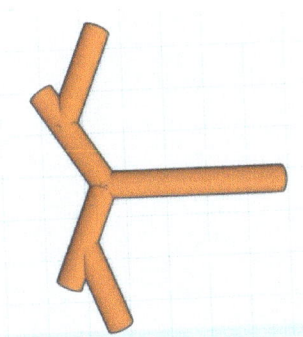

Figure 2: This is the shape of our simple vessel structure. This is not specific to the hepatic portal vein, but is a test whether the r3bEL will be able to print it horizontally.

We then sent these designs to the program Slic3r, which cut the designs into thin layers for the printer to create. These layers were sent to the program Pronterface, then sent instructions to the r3bEL to print the structure.

Results Design One

The r3bEL was unable to accurately print our design (**Pic. 4**). Although the r3bEL was able to accurately lay the first layer of the vein, the second layer could not adhere to the previous layer because it was essentially placing the Pluronic F127 on air. The r3bEL printer couldn't print supports for the design in order to make a cylindrical shape, so it left many blank spots and non-smooth areas on the structure. Also, the printer bed was warped when it was designed, so it was unlevel. This was an issue that we had to fix.

Picture 4: This is the result of the printing of our first design. There are many holes in the print and the print is not smooth due to lack of supports for the print.

Design Two

After fixing the technology issues with design one, we decided to test print a portion of the hepatic portal vein with a design from the National Institute of Health 3D Print Exchange (**Fig. 3**).[3] The hepatic portal vein is the small main portion of the entire biliary structure. We attempted to print only this small section without the individual branches.

At the entrance to the liver, the hepatic portal vein is approximately 11.545 ± 1.514 mm.[8] We printed it at a scale of ⅙ times the original vessel, or 1.924 ± 0.252. The

smaller branches are approximately 6 mm in diameter, so we scaled this area down to 1 mm. Those areas are labeled in the picture above (**Fig. 3**).

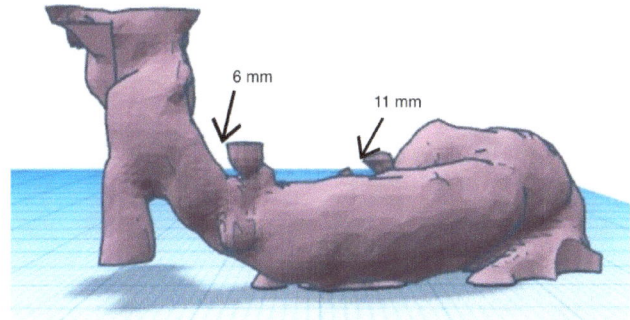

Figure 3: This is an image of the model that was created, along with arrows labeling diameters of the hepatic portal vein section.

Results Design Two

Despite our efforts to print the hepatic portal vein, printing cylindrical structures horizontally still proved to be very difficult when using Pluronic F127 and the r3bEL 3D Bioprinter. As seen in **Pic. 5**, the vertical portion of the design did not print properly and collapsed. This can be seen on the left side of the structure in **Fig. 3**, and the right side of the print in **Pic. 5**. The r3bEL Bioprinter was unable to print that section of the design due to the absence of support underneath that section.

Picture 5: These are the results of the horizontal print attempts. The r3bEL was unable to print an accurate horizontal structure due to the lack of supports.

Pic. 5 reveals flaws in the horizontal printing of the vessel structure, compared to the design seen in **Fig. 3**. This is because the r3bEL lacks the ability to print using supports. Due to the extrusion based technology and the inability to print supports for our cylindrical designs, we decided to try a new method to print vessel structures.

Design Three

Since printing accurate cylindrical shapes horizontally was shown in multiple models to not be possible with the r3bEL 3D Bioprinter, we designed an innovative new way to approach 3D printing of blood vessel structures.

Instead of printing cylinders horizontally, we thought that vertically printing cylindrical structures would better maintain their shape. They can later be placed together to form the entire vessel structure. To test this idea, we decided to print vessels of various diameters and measure the height at which they begin to print inaccurately.

To do this, we designed five different cylindrical structures similar to the portal vein in diameter, and printed the structures, measuring height as they were printed. The

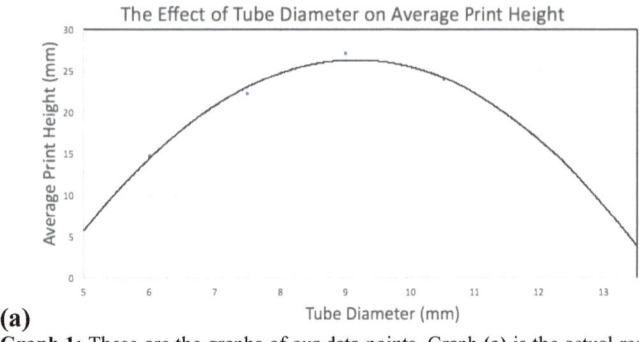
(a)

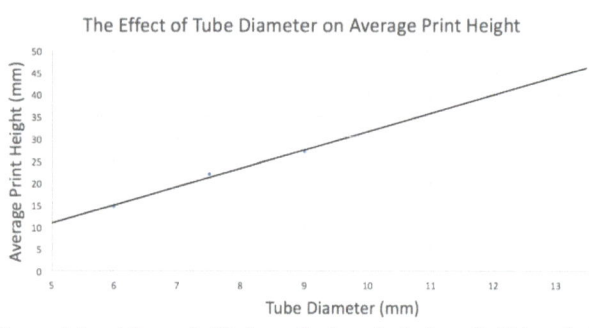
(b)

Graph 1: These are the graphs of our data points. Graph **(a)** is the actual results of our prints, while graph **(b)** shows the hypothetical results if the printing method had been constant for all cylinders.

diameters used were 6 mm, 7.5 mm, 9 mm, and 10.5 mm, since the portal vein ranges from 6 to about 11 mm **(Fig. 4)**. Each of these were printed 20 times using Pluronic F127.

Results Design Three

The 6 mm cylinder printed to an average height of 14.7 mm **(Pic. 6)**. Once each cylinder reached its maximum height, it began to print inaccurately at the top, creating a pool of Pluronic F127 that did not line up with the edges of the cylinder already printed **(Pic. 6)**. As the cylinders got taller, they began to wobble back and forth, which resulted in the inaccuracy of the prints.

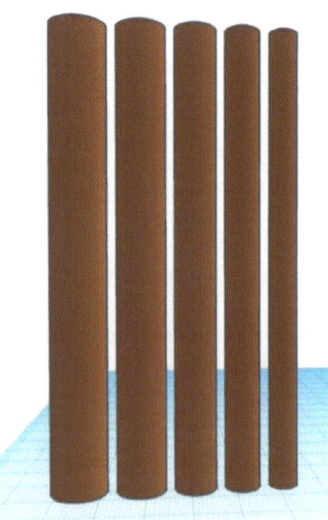

Figure 4: These are the five different cylinders that we designed to print vertically using the r3bEL. They are of diameters 10.5, 9, 7.5, and 6 mm from left to right.

(a)

(b)

Picture 6: (a) Vertical cylinder printed with a diameter of 6 mm, the smallest diameter of the portal vein. **(b)** Once the cylinder reached its maximum height, it would print inaccurately at the top of the cylinder.

The 7.5 mm diameter cylinder printed to an average height of 22.15 mm before printing inaccurately and the 9 mm cylinder printed to an average height of 27.15 mm. When we reached 10.5 mm or larger diameter cylinders, the printer began to print in a different pattern than the pattern for the smaller prints. The prints weren't as accurate or effective. The 10.5 mm diameter cylinders reached 23.85 mm on average. In addition, the sides of the cylinders were not as smooth as the shorter cylinders **(Pic. 6 and 7)**.

When the Slic3r program created the printing instructions for the 6, 7.5, and 9 mm diameter cylinders, all were programmed with a fill pattern that differed from the 10.5 mm. The pre-selected fill pattern in Slic3r was hexagonal in shape. The cylinders with diameters under 10 mm were actually too small to fit hexagons, but once the diameter was 10 mm or above, it was able to fit the hexagons **(Fig. 5)**. We believe that this is the reason the prints looked rough and uneven and were much less effective.

Picture 7: This is the outcome of the 10.5 mm print. As seen, the sides are more rough on this print than the previous ones because of the different fill pattern.

The average height at which each of these cylinders started to print inaccurately, along with a line of best fit for these heights, was recorded and graphed, shown in **Graph 1**. We also developed an equation that would predict average cylinder height based on diameter using the r3bEL 3D Bioprinter. This equation is $y = -1.4x^2 + 24.893x - 84.645$. We can estimate the height of larger diameter prints using this equation and graph our prediction. Using the same print method for all structures, we can predict that the 10.5 mm cylinder would have printed to a height of approximately 33.78 mm.

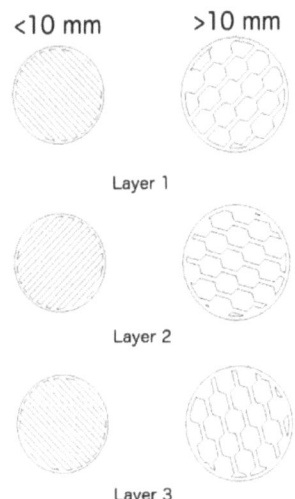

Layer 1

Layer 2

Layer 3

Figure 5: These are the different fill patterns for the prints less than 10 mm in diameter and greater than 10 mm in diameter. Both settings were the same, but because diameters greater than 10 mm have a larger area, the hexagons fit. The printer printed the layers in order of Layer 1, Layer 2, Layer 3, then repeated that order for the remainder of the print.

DISCUSSION

3D bioprinting is a new and exciting method for the future of organ transplants. With the ability to create the vascularization of organs, we would be able to transplant 3D printed structures into human bodies. Initially, our goal of this research was to determine the most effective method to print the hepatic portal vein in the human liver. Ultimately, our research led us to examine a new method for printing vascular structures: vertical printing.

Our research began with designing vascular structures in the CAD software Tinkercad. We designed the hepatic portal vein scaled down so that we could print it using the r3bEL 3D Bioprinter, but the printer had difficulty creating this structure with no supports. Because of this, we worked on printing cylinders of several different diameters and examining how tall they could print before printing inaccurately. Piecing together smaller cylinders would allow for printing of a larger vascular structure. This method of printing could have important applications in the future of bioprinting.

Initially, our cylinders printed higher at larger diameters, with the 6 mm diameter printing to an average of 14.7 mm and the 9 mm diameter printing to an average of 27.15 mm. Although we expected the largest diameter cylinders to print the highest average, we found that the 10.5 mm diameter cylinders only printed to an average height of 23.85 mm. However, an unexpected function of the printer was observed with printing at the larger diameters; the pattern the printer followed to print differed from that of the smaller diameters. The movement of the extruder was also faster and more rigid, following a honeycomb shape instead of a zig zag, and the prints looked apparently less structurally sound than the others. The larger diameter structures weren't built as strong as the others and printed inaccurately even at lower heights. This issue can be attributed to settings that could have been changed in Slic3r. We believe the vessel would have continued to follow the linear growth pattern if the print pattern had remained consistent. This research demonstrates that the vertical print material may be applicable to printing this vein in humans due to the approximate size of the liver, which averages 7 cm wide in women and 10.5 cm wide in men.[16]

We encountered many challenges throughout this research. When we first received the printer, there was an issue with the homing ability, so our mentor sent us a new printer that was properly working. After receiving the working printer, we had problems getting the printer to accurately print our designs due to issues with the build of the printer. The printer bed was unintentionally warped when it was built, which resulted in an unlevel printing surface. This caused the printer to attempt to print our designs in mid-air. Due to these issues, our timeline for conducting research was shortened significantly, and we were unable to alter the programming of the 10.5 mm cylinder to test the more effective print pattern for the larger cylinders. Had we been able to change the program, we believe that the 10.5 mm cylinder would have printed at a height that continued the linear progression.

The next step with vertical printing would be to print entire structures vertically, such as the entire portal vein. While we have not tried to print the entire vein, our data suggests that it may be possible. Printing vertically may require the researcher to piece together several small sections of cylinders to form the larger, more complex structure.

Another limitation of this research is the bio ink that we used. The gel-like Pluronic F127 bioink could not adequately hold its shape when printing circular structures. Instead, biodegradable thermoplastics used by normal 3D printers could be much more effective for creating structures such as the hepatic portal vein. These standard 3D printers also have the ability to print supports which can later be removed. Printing the hepatic portal vein using biodegradable thermoplastics with a traditional 3D printer would enable the printed structure to maintain its shape in a 3D plane, and could then be suspended in a block of bioink, dissolved away, so the remaining hollow space could be seeded with endothelial cells.

The r3bEL 3D Bioprinter was built to print structures with a diameter smaller than 1 cm and was designed to be an educational tool, rather than a professional 3D bioprinter. As such, the promise of 3D bioprinting may be even greater when using higher-end 3D bioprinting equipment. However, our data suggests that the vertical printing method we used could be a viable option in industry when printing vascular structures in the future. The circular structures held their shape well when printed vertically. If pieced together, a complex network of blood vessels could in theory be printed in its entirety.

ACKNOWLEDGMENTS
We would like to thank Dr. Mayasari Lim for her help in mentoring us in this research. She supported us in the design and implementation of our project. We would also like to thank Vignesh Krishna and Sarah Salameh, engineers at SE3D Education, for their help in setting up the r3bEL and teaching us how to use several applications associated with the printer. We would like to thank Jennifer Timmons for funding our research and Bryan Winkelman for helping us set up and maintain our web page, blog, and for managing our incoming and outgoing finances. We

would like to thank Jim McClurg and Amy England for helping to film and edit the video for our webpage and Wendy Lerolland for her editorial assistance and feedback on our research. We would like to thank Tom Dillon for his guidance and suggestions during the experimental design process. We would like to thank Nikki Dobos and Susan Petri for sharing lab space. Lastly, we would like to thank Rock Canyon High School and Douglas County School District for supporting our research and providing the lab space and equipment to conduct our research.

REFERENCES

1. About The Liver. *Arizona Transplants Association, PC*. Retrieved 2017, April 18. [Web]
2. Bertassoni, L. E., Cecconi, M., Manoharan, V., Nikkhah, M., Hjortnaes, J., Cristino, A. L., . . . Khademhosseini, A. (2014). Hydrogel bioprinted microchannel networks for vascularization of tissue engineering constructs. *Lab Chip,14*(13), 2202-2211. DOI:10.1039/c4lc00030g.
3. Dilmen, N. (2015). Biliary system. *National Institute of Health*. Retrieved 2017, April 12. [Web]
4. Hernandez, R., & Brown, D. T. (2010). Growth and maintenance of mosquito cell lines. *Current Protocols in Microbiology*. DOI:10.1002/9780471729259.mca04js17.
5. Internal Anatomy of the Liver. (2010). *Loma Linda University Health*. Retrieved 2016, October 23. [Web]
6. Kolesky, D. B., Truby, R. L., Gladman, A. S., Busbee, T. A., Homan, K. A., & Lewis, J. A. (2014). 3D Bioprinting of vascularized, heterogeneous cell-laden tissue constructs. *Advanced Materials, 26*(19), 3124-3130. DOI:10.1002/adma.201305506.
7. Kuo, C., Eranki, A., Placone, J. K., Rhodes, K. R., Aranda-Espinoza, H., Fernandes, R., . . . Kim, P. C. (2016). Development of a 3D printed, bioengineered placenta model to evaluate the role of trophoblast migration in preeclampsia. *ACS Biomaterials Science & Engineering, 2*(10), 1817-1826. DOI:10.1021/acsbiomaterials.6b00031.
8. Mandal, L., Mandal, S. K., Bandyopadhyay, D., & Datta, S. (2011). Correlation of portal vein diameter and splenic size with gastro-oesophageal varices in cirrhosis of liver. *Indian Academy of Clinical Medicine, 12*(4), 266-270.
9. Mandrycky, C., Wang, Z., Kim, K., & Kim, D. H. (2016). 3D bioprinting for engineering complex tissues. *Biotechnology Advances, 34*(4), 422-434.
10. Mills, A. (2015). Bioprinting in 3D: Looks like candy, could regenerate nerve cells. *Michigan Tech*. Retrieved 2017, February 27. [Web]
11. More About Organ Donation. (2015). *American Liver Foundation*. Retrieved 2016, November 04. [Web]
12. Organ Procurement and Transplantation Network. (2017). *Health Resources and Services Administration*. Retrieved 2017, March 29. [Web]
13. PCL Viability. (2016). *BioBots Inc*. Retrieved 2016, October 23. [Web]
14. Runyon, B. A., Gores, G., & Talwalkar, J. A. (2014). Cirrhosis | *The National Institute of Diabetes and Digestive and Kidney Diseases*. Retrieved 2017, March 31. [Web]
15. What is 3D printing? How does 3D printing work? (2015). *3D Printing.com*. Retrieved 2016, October 24. [Web]
16. Wolf, D. C. (1990). Evaluation of the Size, Shape, and Consistency of the Liver. *National Center for Biotechnology Information*. Retrieved 2017, April 12. [Web]

ABOUT THE AUTHORS

Pictured: From left to right, Sam Martin and Bailey Timmons pictured with their mentor Dr. Mayasari Lim with SE3D Education and the r3bEL 3D Bioprinter.

Throughout our research, we discovered a lot of what science is like in the real world, as opposed to the school environment. Problems rose at every step of our research. We couldn't figure out a way to design an accurate model of the hepatic portal vein, which ended up taking a lot more time than we originally thought it would take. The first printer we received was broken, which meant we did not have the printer ready until about two months into our research. Near the end of our research, we learned from our mentor that printing horizontal would be impossible using our selected bioink, which was a huge setback. We couldn't give up on all the progress we had made, so we decided to explore new methods that we hoped could be useful for future groups. We have never faced so many setbacks in any previous class, and we have gained a better understanding of what our future as scientists/engineers could look like.

In a typical science class, students learn about how the body works or how chemicals react together, but instead we learned what it's like to write professional emails, how scientific writing differs from creative, how to understand a scientific journal article and write our own, why things take so long in the scientific world, and why getting clear results from scientific research is so difficult. Students in our class have been told that high schoolers can't conduct novel research, but we have and next year's biotech students will do the same. This program provides an opportunity like no other for students.

Comparing the relative abundance of environmental nontuberculous mycobacteria between Colorado and Hawai'i

S. S. Narayan, O. H. Voss, and S. L. Fordham
Department of Science, Principles of Experimental Design in Biotechnology, Rock Canyon High School, Highlands Ranch, Colorado, USA

Environmental nontuberculous mycobacteria (NTM) are found ubiquitously in water and soil. NTM is heavily implicated in pulmonary diseases as a result of infection; contributing factors for infection vary, but Hawai'i has the highest instance of infection in the United States. We tested the relative abundance of NTM between Colorado and Hawai'i with data gathered from microbial growth on culture plates, water quality assays, and microbiome sequencing. While we had initially planned to use the microbiome sequence data and environmental data to draw comparisons between the two states, only the Colorado samples were able to be sequenced and analyzed. We were able to identify overall differences between the two states with the following environmental assays: pH, zinc, iron, water hardness, nitrite, temperature, and humidity. Additionally, we observed phenotypic differences in the microbial growth on the bacterial culture plates between Colorado and Hawai'i. The Colorado 7H10 culture plates grew a fast-growing, pale yellow species of mycobacteria which was later confirmed as *M. gilvum* through sequencing. The Hawai'i 7H10 culture plates, in contrast, were more diverse in the number of species, and included dark brown, orange, and white colonies. Since none of the Hawai'i samples resulted in successful sequence data, we were unable to identify the species of mycobacteria present on these culture plates. It is important to note that *M. avium*, the species of mycobacteria that is commonly associated with NTM lung infection, was not found in any of the sequence data we obtained from Colorado. Based on microbiome sequencing of the Colorado samples, we found that the mycobacteria biodiversity and its proportion in the sample was much higher in December than in January. In December, the air temperature was 11.04 °C warmer than in January and the relative humidity was 22.2% lower. Because there are significant differences in temperature and humidity between Hawai'i and Colorado, it would be important to further investigate the effects of these factors and of pH, nitrite, water hardness, zinc, and iron on NTM relative abundance.

The genus *Mycobacteria* contains a plethora of species, including nontuberculous mycobacteria (NTM), which naturally inhabit water and soil. There are more than 174 NTM species identified to date, with an additional level of subclassification, slow-growing and fast-growing mycobacteria.[18] Commonly found mycobacterial pathogenic isolates include *Mycobacterium tuberculosis, Mycobacterium avium, Mycobacterium intracellulare, and Mycobacterium chimaera.*[11]

As naturally occurring environmental organisms, these bacteria contribute to NTM pulmonary diseases (NTM-PD) found in humans.[4] NTM-PD differs from its mycobacterial cousin, tuberculosis, as it is not currently treatable with antibiotics and is not life-threatening.[4,17] NTM-PD manifests itself through symptoms such as a recurring cough, sweating, fever, weight loss, loss of appetite, and chronic fatigue.[3] It is most commonly acquired through inhalation of NTM in the aerosolized form. Due to NTM's ubiquity, it can be supposed that everyone has, at some point, been exposed to or inhaled NTM. Exposure, however, does not directly result in NTM infection. *M. avium*, one of the aforementioned species, has been implicated in close to 90% of NTM-PD cases, specifically the pathogenic isolates from house plumbing.[4]

Though common in households, contracting NTM-PD from plumbing is only one of the many ways for NTM-PD to manifest, as NTM is found in soil and can be contracted in many other situations. In addition, a number of other factors correlate NTM exposure to lung disease. Conditions that predispose people to develop NTM-PD include scoliosis, prior lung-related disease, and immunodeficiency. Otherwise, Asian-Pacific Islanders have been shown to have the highest incidence of NTM infection.[11]

In the United States, Hawai'i has the highest number of NTM-PD cases: 396 per 100,000 people.[2] By comparison, Colorado has a mean of 1.7 cases per 100,000 people.[13] The high number of NTM cases in Hawai'i may be due to the humid and tropical climate which is a conducive environment for bacteria to grow. Colorado, in comparison, is quite dry and has high altitude, which hinders bacterial growth.[2] Using these environmental differences, we sought to use our research to answer whether environmental characteristics played a role in the NTM case rate.

To date, little research has been conducted on NTM even though NTM-PD rates are increasing,[5] leaving a significant gap in our understanding of NTM. Conducting research on the relative abundance of NTM in both Hawai'i and Colorado will contribute to the growing body of information about NTM, and help reveal the correlations between the locations and the NTM case rate.

Drawing on a previous study conducted by our mentor, Dr. Jennifer Honda with National Jewish Health, on the biofilms of showerheads in both Colorado and Hawai'i, we studied the relationship of NTM between Colorado and Hawai'i. This previous study, among others, showed that Hawai'i had a higher abundance of NTM than Colorado.[1] Thus, we collected samples in both locations, using the Colorado samples as a baseline for our data.[1] In addition to the comparison of NTM abundance in both Colorado and Hawai'i, we also examined whether various environmental factors contribute to the NTM infection rate disparity, including pH, chlorine, nitrite, nitrate, zinc, iron, alkalinity, and water hardness.

METHODS

Study Design

In this study, we compared NTM prevalence between different water sources in both Colorado and Hawai'i. We sampled in Colorado and our partners from the Iolani School sampled in Hawai'i using a common swabbing procedure, which will later be detailed. Sampling was conducted in both states on December 4, 2016 and January 7, 2017. Sites were chosen to correspond to one another between the states in order for apt comparison to be run between them.

Map 1: The Colorado sample sites (Centennial Water Treatment Plant, Last Chance Ditch, South Platte River, McLellan Reservoir, Rock Canyon, and the Well) are shown above.

The South Platte River (39.5713°N, 105.0513°W) (SP) and Last Chance Ditch (39.5757°N, 105.0465°W) (LCD) sample location in Colorado will be compared with the Ala Wai Canal (AWC) (21.2851°N, 157.8285°W) in Hawai'i. The McLellan Reservoir (MCL) (39.5713°N, 105.0513°W) sample in Colorado will be compared with the Ala Wai Boat Harbor (H) in Hawai'i (21.2853°N, 157.8401°W). The well water sample (WW) at a Castle Pines home in Colorado (39.4829°N, 104.8895°W) and the surface water sample from the Centennial Water Treatment Plant (WTP) in Colorado (39.5601°N, 105.0198°W) will both be compared to the sample collected at the Princess K Well (PK) in Hawai'i (21.2775°N, 157.8259°W). Lastly, the sample of the faucets collected in the girls' locker room at Rock Canyon High School (39.5200°N, 104.9215°W) (RC)

will be compared with the sample of a bathroom sink and the water fountain sample at the Iolani School (Io-S) in Hawai'i (21.2854°N, 157.8244°W) **(Map 1, Map 2, Pic. 1)**.

Map 2: Shown above are the Hawai'i sample sites: the Ala Wai Harbor, the Ala Wai Canal, the Princess Kei Well, and the Iolani School.

Picture 1: These photos depict every sample location between both dates. Comparable locations between Colorado (CO) and Hawai'i (HI) are organized in the same row with columns showing the different conditions (if applicable) associated with the two samples.

Sample Collection

Due to NTM's hydrophobic nature, swabbing was done on the surfaces of rocks that were in contact with but not submerged in the water. Similarly, in locations where rocks were not accessible, pipes, faucets, and docks were swabbed instead. Out of the three samples collected at every location, one swab was used for DNA extraction, while the other two were used to streak two different types of culture plates.

DNA Extraction for Microbiome Sequencing

DNA was then extracted from one of the three swabs collected at each sample location using the PowerSoil DNA Isolation Kit from MoBio. The protocols detailed by the kit were followed with alterations made for the initial steps by our mentor, Dr. Jennifer Honda. Prior to performing the extraction, we added 1 mL of sterile water into sterile Eppendorf tubes before placing the broken-off swab in the water. We then vortexed the samples for one minute to remove microbial samples from the swab. Once this was complete, we added 600 uL of the water mixture into the PowerBead Tubes and added 60 uL of C1 solution. After this step, the protocols were followed as outlined in the kit. The DNA samples were then stored at -20°C until sequencing could be performed. We then confirmed DNA concentration for each sample by using a nanodrop lite. The samples were then sent to Mr. DNA for microbiome sequencing. Once the sequencing was completed, the relative abundance of bacteria was analyzed by Dr. Jessica Joyner and Dr. Theodore Muth at the City University of New York (CUNY).

Microbial Culturing Plates

The remaining two swabs from each location were used to streak bacterial plates for culture. One swab was used to streak a 100 mm TSA petri plate, while the other was used to streak a 100 mm 7H10 petri plate. The TSA media is nonspecific and allows bacterial growth of all kinds, while the 7H10 plate is specific to only NTM bacterial species. This resulted in a total of four culture plates for each location; two for each sampling date.

Colorado bacterial plates were streaked in the biological safety cabinet of Rock Canyon's lab, following sterile protocols (**Pic. 3**). They were incubated at 37°C for 2 days, after which they were stored at 4°C to prevent overgrowth. After we photographed the plates, they were given to our mentor, Dr. Jennifer Honda for further analysis. The Hawai'i plates were prepared following similar protocols for culturing and incubation at the Iolani High School.

Environmental Factors and Water Profile Assays

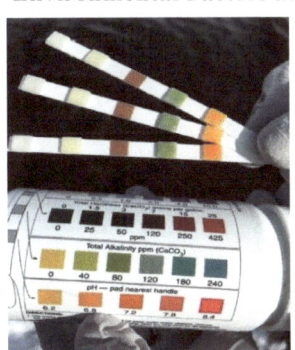

Picture 2: Depicted above is an image of the dipstick assay with corresponding color scales for pH, alkalinity, and total hardness.

To draw comparisons between samples and sample locations, dipstick tests were used to test the water at each sample site. The water was tested for pH, chlorine, nitrite, nitrate, zinc, iron, alkalinity, and water hardness. The results of these quantitative assays were compared between each sample location, as well as between the two different sampling dates.

The dipstick assays used were WaterWorks ZincCheck Dipsticks, Aquacheck Water Quality 5 in 1 Strips, and Aquacheck Water Quality Nitrate/Nitrite Strips. All tests were performed exactly as

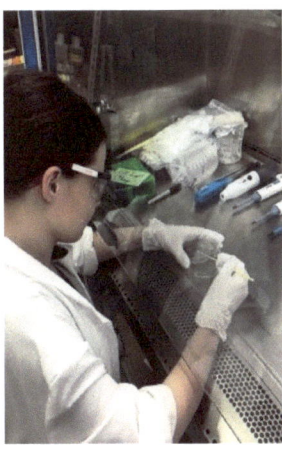

Picture 3: Pictured above is Voss streaking the culture plates with a bacterial loop in the biological safety cabinet of Rock Canyon's Lab.

indicated in the instructions for each test and subsequently quantified using the provided color scale.

RESULTS

We collected samples at various locations in both Colorado and Hawai'i in order to better understand the relationship between NTM and the environment. We approached this comparison by collecting environmental and microbial data. Our microbial data assays included culturing bacteria on TSA and 7H10 culture plates in addition to performing microbiome sequencing on each sample.

The abbreviations for the sample sites on the graphs below are as follows: in Colorado, MCL for the McLellan Reservoir, SP for the South Platte River, LCD for the Last Chance Ditch, WTP for the Centennial Water Treatment Plant, RC for Rock Canyon, WW for Well Water; and in Hawai'i, PK for Princess Kei Well, AWC for Ala Wai Canal, Io-S for Iolani Sink, Io-F for Iolani Fountain, and H for Harbor.

Environmental Data

Air temperature, humidity, pH, alkalinity, water hardness, zinc, iron, chlorine, nitrite, and nitrate were all tested at each location. Several of these were noticeably different between the sampling sites in each state.

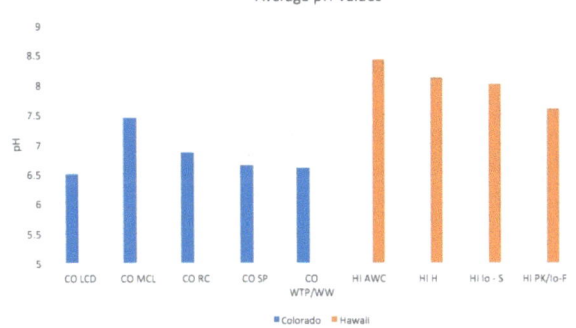

Graph 1: The graph above shows the average pH levels for each sample site in Colorado and Hawai'i. The pH at each site is the average of three tests taken on each sample date in both December and January.

The pH differs noticeably between Colorado and Hawai'i with the pH of the water at the Colorado sample sites more than 10 times more acidic on average compared to the Hawai'i samples (**Graph 1, Pic. 2**).

The zinc level in the water was a second environmental factor in which we found a significant difference between Hawai'i and Colorado. Zinc was detected at every sample in Colorado but was undetectable to the test strips at all but the Iolani High School in Hawai'i which was sampled twice.

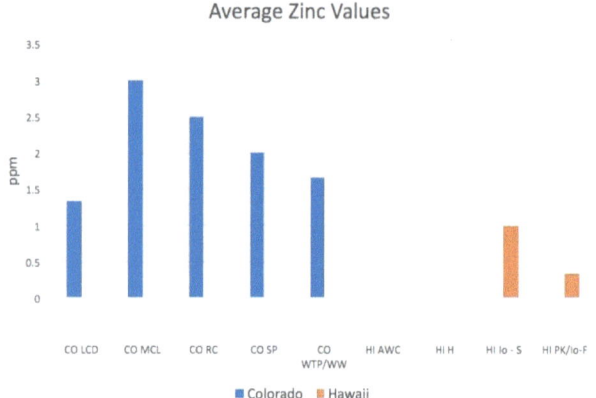

Graph 2: The graph above shows the average zinc levels for each sample site in Colorado and Hawai'i. The zinc at each site is the average of three tests taken on each sample date in both December and January.

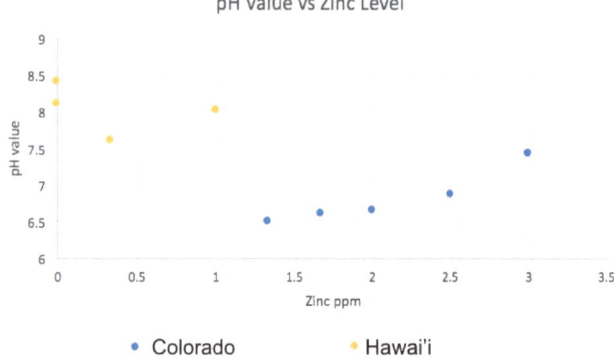

Graph 3: The graph above shows the relationship between zinc and pH levels in Colorado and Hawai'i. Each dot represents the average of that site between the two sample dates.

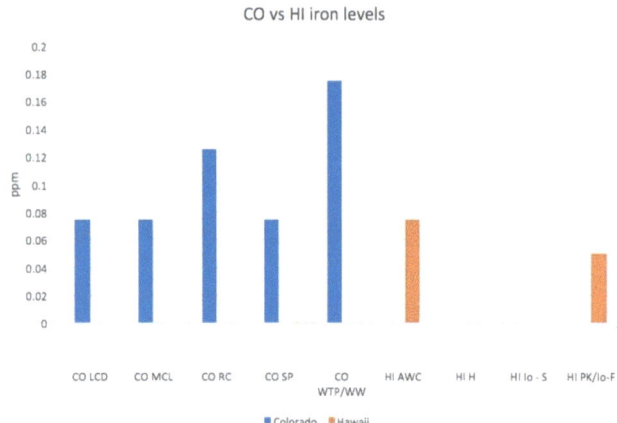

Graph 4: The graph above shows the average iron levels for each sample site in Colorado and Hawai'i. The iron at each site is the average of three tests taken on each sample date in both December and January.

Similarly, iron was also detected at every sample site in Colorado, at levels higher than what is recommended by the EPA, and in only half of the Hawai'i samples (**Graph 4**).[6,20]

One area of particular interest when studying microbial growth is temperature and humidity. Many bacterial species grow at very specific temperatures and often require higher humidity to survive in the environment. Not surprisingly, the air temperatures and humidity levels were significantly

higher in Hawai'i compared to those in Colorado during the winter months of our experiment.

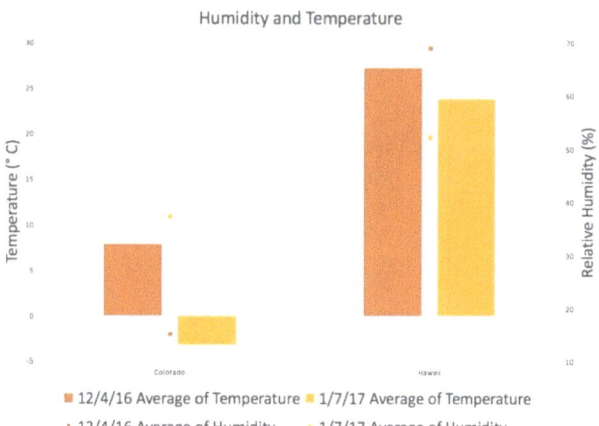

Graph 5: The graph above shows the comparison and relationship between the air temperature and relative humidity on December 4, 2016 and January 7, 2017 in Colorado and Hawai'i.

The other environmental data, including nitrate, nitrite, water hardness, and alkalinity did not show any discernable patterns between the Colorado and Hawai'i samples. However, there were a few notable outliers within our environmental assays. First, chlorine was only detectable in the Centennial Water Treatment Plant sample, which is significant because chlorine is a known antimicrobial agent. Likewise, nitrite was only detected at the McLellan Reservoir in December and was 10 times the EPA limit for public water systems at a value of 10 ppm.[16] No other nitrite levels were detectable at any Colorado sample location on any other date.

While the water hardness did not appear to have a clear pattern, upon further examination, there was an outlier in the sequence data we collected for Colorado. The majority of the Colorado samples, regardless of date, had a hardness of either 120 or 250 ppm, while the Rock Canyon sample in December spiked at 425 ppm. This higher value was also observed at several of the Hawai'i sample sites, which were consistently measured as either 250 or 425 ppm with one outlier of 120 ppm. The significance of this data will be discussed further in the microbiome section.

Microbial Culture Data

Bacterial culture data served as an initial indicator in understanding the bacterial presence at each sample site, and allowed initial comparisons to be made in tandem with the environmental characteristics and microbiome sequencing. It also allowed us to draw comparisons from sample sites that were not sequenced.

All the Colorado TSA plates, save for the water treatment plant, had extreme growth. The water treatment plant culture plate had no type of bacterial growth on either plate, and was the only sample where chlorine was detected.

Growth on the 7H10 plates in Colorado displayed two kinds of growth: lawns and spotted colonies. It is important to acknowledge the phenotypical differences between the Hawai'i and Colorado colonies to further understand the significance of the culture data. To begin with, the coloration of bacteria was very different between the two locations: Colorado tended to have overall bright yellow bacterial growth on the TSA plates and a pale yellow to nearly white population of mycobacteria on the 7H10 plates. In contrast, the bacterial colonies from the Hawai'i samples were overall more diverse with several phenotypically different colonies growing on the TSA and 7H10 plates. No bright yellow colonies were observed and the 7H10 plates had white, orange, and dark brown colonies present (**Pic. 4**).

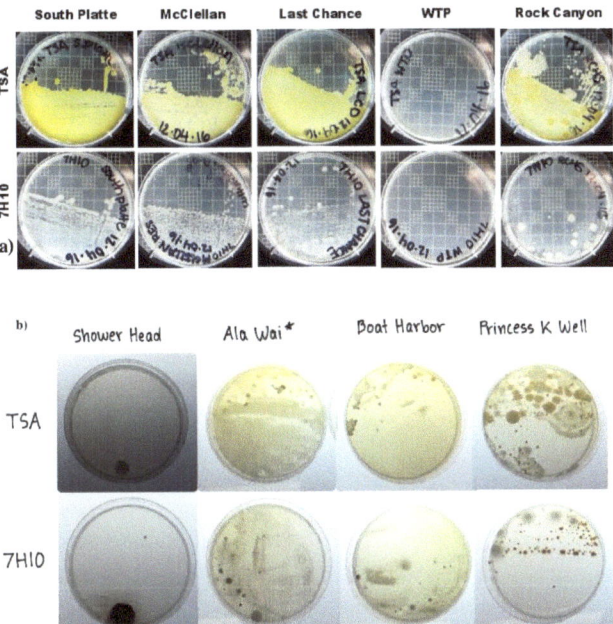

Picture 4: These pictures show the bacterial culture plates that were streaked in December at each sample location. The TSA plates include all microbial growth while the 7H10 plates are specific to mycobacteria. (a) These are the culture plates from the Colorado sites and (b) these are the culture plates from the Hawai'i sites.

Microbiome Sequencing

We sent a total of 18 samples to Mr. DNA, and of these samples, none of the Hawai'i samples were able to be sequenced and analyzed due to either poor quality DNA and/or low sequence numbers, which could be attributed to the issues with shipping that we had encountered. The Hawai'i samples were delayed for an entire week prior to arriving in Colorado. We do not know the conditions in which the samples were stored during this time. In addition, three Colorado samples were unable to be sequenced. One of these was the water treatment plant sample, the only sample with a detectable amount of chlorine. This sample may have failed to produce sequence data due to the lack of bacteria present in these samples. In addition, the well water and the January McLellan Reservoir samples did not result in sequence data that could be analyzed.

With the sequence data we did have, we were able to compare the differences in mycobacteria content between the Colorado samples from December and January. Rock Canyon in December had the largest proportion of mycobacteria within the sample compared to the other

Colorado sample sites, but it dropped significantly in January (**Graph 6**). In our environmental assays, we noted that the water hardness was the highest in December at Rock Canyon and was much lower in January at the same sampling site. This correlates with the proportion of mycobacterium found at this location. In addition, nitrite was only detectable at the McLellan Reservoir in December, which had the lowest proportion of mycobacterium out of all the December samples.

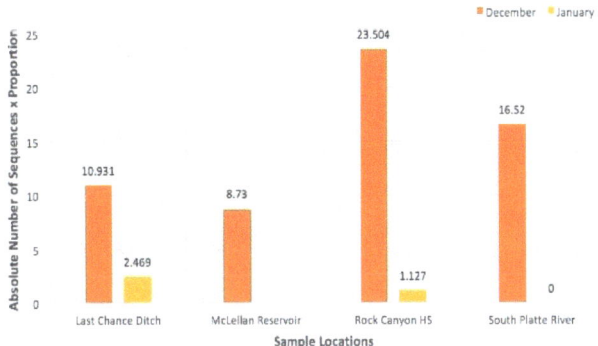

Graph 6: The graph shows the absolute number of mycobacteria within each sample. The McLellan Reservoir was not able to be sequenced in January.

The most significant environmental factors that correlate with our mycobacterium proportion data were air temperature and relative humidity. The temperature in December was much higher than the January temperatures, which were below the freezing point. The proportion of mycobacterium in the January samples were statistically lower than in the December samples within a 95% confidence interval. The average absolute number of mycobacteria in December was 14.92 while in January it was 1.199. January had higher relative humidity, ranging from 35 to 40%, whereas the December sample sites measured at only 13-15%. The last variable environmental factor to consider is the surface that was swabbed when collecting the sample. At Rock Canyon High School, a faucet was swabbed, compared to the rock surfaces swabbed at all other locations.

We also created a heatmap to show the differences in the biodiversity of the mycobacterium species present within the samples between December and January in Colorado (**Fig. 1**). Though we are characterizing biodiversity by the number of bands on the heatmap, this does not necessarily denote a species, but is indicative of a unique DNA sequence.

The Rock Canyon sample in December had the highest overall biodiversity within the sample as indicated by the presence of five distinct bands, whereas in January, the Rock Canyon Sample was considerably lower in both biodiversity and abundance with only one dark blue band present. Overall, the January samples exhibited much less biodiversity of sequences than their December counterparts.

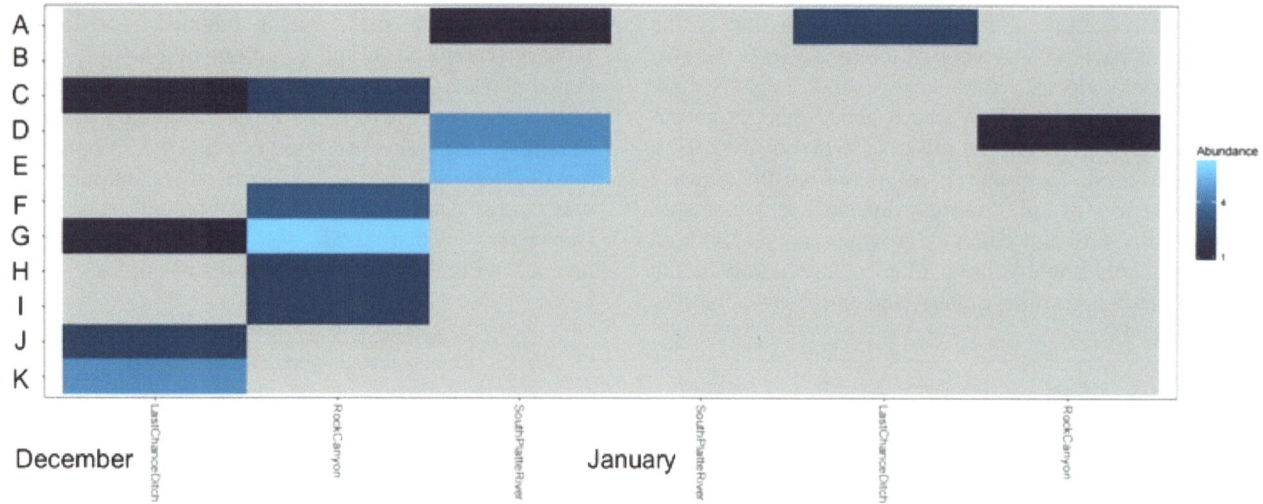

Figure 1: The heatmap above portrays the abundance of mycobacteria subspecies at each Colorado sample location, and by extension displays the biodiversity of mycobacteria in a sample. Each band represents a different sequence and the abundance of that sequence in the sample is indicated by the shade of blue.

The abundance of these sequences were also much lower. The abundance data from this heat map matches the data in **Graph 5**.

The most abundant species overall was found in the December Rock Canyon sample and identified as *M. gilvum* **(Point G, Fig. 1).** This species was also found at Last Chance Ditch in December but at a much lower abundance. We also identified either *M. fortuitum* or *M. septicum* as the two species present in January; they were also present in December. The South Platte River sample didn't have any detectable mycobacterial growth in January. This is important to note considering the proximity and relation of the Last Chance Ditch to the South Platte River; the former feeds raw, untreated water into the South Platte River.

DISCUSSION

Nontuberculous mycobacteria (NTM) remains relatively understudied, and its relationship with the environment is still unclear. The main purpose of this research was to identify whether the following environmental factors played a role in NTM presence: air temperature, humidity, pH, alkalinity, water hardness, zinc, iron, chlorine, nitrite, and nitrate. To examine this relationship, we compared Colorado and Hawai'i using three types of data: environmental assays, bacterial culture growth, and microbiome sequencing. Though we had initially wanted to use the microbiome sequence data to compare Colorado and Hawai'i, the Hawai'i samples were unable to be sequenced and analyzed. Due to this, the comparison between the states was done primarily using the environmental assays.

There were discernable differences between the pH measured at the sites in Colorado and Hawai'i. Mycobacteria show preference for lower to more neutral pH levels, which were present in Colorado compared to the more basic levels found in Hawai'i.[6] Additionally, Hawai'i's climate is characterized by both high temperatures and high relative humidity levels; many researchers have assumed this to be one of the major reasons why the NTM case rate in Hawai'i is so high.

While we don't have sequence data comparing Hawai'i to Colorado, we did find a statistically higher relative abundance of mycobacteria in the Colorado samples from December, which was a much warmer but less humid day. In January, the water was frozen over and all the rocks were covered in a thin layer of snow. From the microbial culture data, we observed a difference in the overall diversity of mycobacteria on the 7H10 bacterial plates between Colorado and Hawai'i. The Hawai'i samples had a wider variety of different types of colonies including non-pigmented, yellow, orange, and brown **(Pic. 4).** Meanwhile, the Colorado samples only had the one colony type, which was pale yellow. While temperature did not inhibit growth of all mycobacteria, it may have impacted the overall diversity of species present. It has been found that different mycobacteria species have different ideal growth rates ranging from 30°C to 37°C and prefer water temperatures of less than 50°C. [8,14]

Two different heavy metals, iron and zinc, were present in all of the water samples in Colorado whereas in Hawai'i the only detectable level of zinc was found in the high school. Iron was found in half of the Hawai'i samples. The iron levels measured in the Colorado samples were higher than the EPA's suggested value of 0.3 mg/L.[6,20] In a previous study of mycobacteria isolates, it was found that *M. avium*, the species of mycobacteria most associated with occurrences of NTM-PD, is rarely found in waters with high concentrations of heavy metals.[9] In all of the samples sequenced in Colorado, *M. gilvum*, *M. fortuitum*, and *M. septicum* were identified but no *M. avium* was found.

While some of the remaining environmental assays did not produce any consistent trends between Colorado and Hawai'i, certain outliers are noteworthy. There was no microbial growth at the water treatment plant, which may have been due to the presence of chlorine. The McLellan Reservoir in December was the only sample that tested positive for nitrite, and was found at levels 10 times the EPA recommended value.[16] This sample also had lowest proportion of mycobacteria. Previous studies have shown nitrite to be an antimicrobial agent for biofilm formation.[22]

In addition, the December Rock Canyon High School sample not only had the highest overall abundance of mycobacteria within the sample, but also had the highest biodiversity of all of the Colorado sites for both dates. This sample was collected on plumbing rather than rocks: plumbing is where mycobacteria most commonly occur.[4,8] It is also important to note that the total water hardness measured at this site was the highest level detected, of all of the Colorado samples, at 425 ppm. In Hawai'i, the water hardness assays also measured 425 ppm at two of the sample sites.

Bacterial growth on the 7H10 plates were used to draw phenotypic comparisons between the sample sites. Differences in color and speed of growth were most noticeable. In December, the Colorado plates grew rapidly, whereas in January they grew more slowly. Additionally, we observed that there was a difference in coloration and overall diversity on the plates. The Colorado plates appeared to have one to two different species of mycobacteria growing that were pale yellow in color, whereas the Hawaii plates had a wider range of diversity of three to four different species that were clear, light yellow, orange, and brown **(Pic. 5)**.

When we BLASTed the mycobacteria sequence data, we were able to identify many species that contribute to the biodiversity of each sample in Colorado. We discovered three main species of mycobacterium found in the samples. *M. gilvum* was the most abundant species overall and was found in December. It is a pale yellow, fast growing NTM, which supports our observation of the December Colorado culture plates.[8] Two species present even with the cold temperatures of January in Colorado were *M. fortuitum* and *M. septicum*. They are both fast-growing strains of NTM: *M. fortuitum* does not cause pulmonary issues, but can lead to infections of the skin, lymph nodes, joints, and osteomyelitis in those who are immunocompromised,[7,15] while *M. septicum* has been implicated in sepsis cases.[12]

There were several limitations and errors associated with this research. The samples collected in Hawai'i were performed by students from the Iolani High School, and variations in sampling protocol may have affected the microbial culture plate growth and sample quality. Additionally, long shipping times of the Hawai'i samples to Colorado as well the samples from Rock Canyon to Mr. DNA might have caused the problems we encountered with the sequencing of the DNA samples.

Our research demonstrates the need for further investigation into the correlation between water temperature, pH, hardness and the presence of heavy metals and nitrite with NTM abundance. In order to expand on this research, better sequence data needs to be collected from both Hawai'i and Colorado along with these environmental assays. Direct assays of these environmental factors with *M. avium* should also be performed specifically.

ACKNOWLEDGMENTS
We would like to especially thank our mentor Dr. Jennifer Honda, a microbiologist at National Jewish Health, for guiding us throughout this research. She has played a significant role in our understanding of NTM and without her support and guidance, implementation of this research would not have been possible. We would also like to thank those who helped fund our research: Dr. Jennifer Honda, Lisa Voss, Sujatha Narayan, Cynthia Martin, Jeanette Ruff, and especially Dr. Ed Chan. We would also like to thank Mr. John Droullard and Patrick Esparza at the Centennial Water Plant, who assisted with our sample collection and helped us understand the water system in Colorado. We are also extremely grateful for Megan and Lynn Baer for graciously allowing us into their home to sample their well, which made our well water comparison possible. We would also like to thank Dr. Yvonne Chan, and her students Emily Kapins and Daisy Chang, for obtaining the Hawai'i data. We would also like to extend our gratitude towards Dr. Jessica Joyner and Dr. Theodore Muth with Authentic Research Experience in Microbiology (AREM) at City University of New York (CUNY) for performing the microbiome analysis and providing funding for some of our samples and Dr. Scot Dowd with Mr. DNA for discounting our sequencing. Additionally, we are extremely grateful towards Dr. Anjali Vaidya, with Front Range Community College, and Drs. Ed Chan, Elaine Epperson, and Victor Ozols with National Jewish Health for donating the MoBio Powersoil Kits. The faculty at Rock Canyon has also been incredibly helpful in this research as well, particularly Bryan Winkelman for designing our website, managing funding, and formatting, Wendy Lerolland for editorial assistance and Nikki Dobos and Susanne Petri for sharing lab space. We extend our gratitude to Tom Dillon, for advising and guiding our project's design. We would also like to thank Amy England and Jim McClurg for filming and editing our research video. Lastly, we would like to thank Rock Canyon High School and the Douglas County School District for providing this opportunity, as well as lab space and the equipment needed for our research.

REFERENCES
1. Abe, J., Alop-Mabuti, A., Burger, P., Button, J., Ellsberry, M., Hitzeman, J., ... Honda, J. R. (2016). Comparing the temporal colonization and microbial diversity of showerhead biofilms in Colorado and Hawai'i. *FEMS Microbiology Letters*, *363*(4), fnw005. DOI:10.1093/femsle/fnw005
2. Adjemian, J., Olivier, K. N., Seitz, A. E., Falkinham, J. O., Holland, S. M., & Prevots, D. R. (2012). Spatial clusters of nontuberculous mycobacterial lung disease in the United States. *American Journal of Respiratory and Critical Care Medicine*, *186*(6),553–558. DOI:10.1164/rccm.201205-0913oc
3. Aksamit, T. R., Falkinham III, J. O., Griffith, D. E., Huitt, G. A., Iseman, M. D., Mitchell, J. D., ... Winthrop, K. (2014). "Insight": A Patient's Perspective. *Nontuberculous Mycobacteria Info & Research,* [Pamphlet].
4. Aksamit, T. R., Philley, J. V., & Griffith, D. E. (2014). Nontuberculous mycobacterial (NTM) lung disease: The top ten essentials. *Respiratory Medicine*, *108*(3), 417–425. DOI:10.1016/j.rmed.2013.09.014
5. Center for Disease Control. Decrease in reported tuberculosis cases—United States, 2009. *Morbidity and Mortality Weekly Report* 2010; 59:289–94.
6. Drinking Water Regulations and Contaminants. (2017). Retrieved 2017, April 15. [Web] Environmental Protection Agency
7. Falkinham, III, J.O. (2009), Surrounded by mycobacteria: nontuberculous mycobacteria in the human environment. *Journal of*

Applied Microbiology, 107:356–367. DOI:10.1111/j.1365-2672.2009.04161.x

8. Falkinham, J. O. (2011). Nontuberculous mycobacteria from household plumbing of patients with nontuberculous mycobacteria disease. *Emerging Infectious Diseases*, 17(3), 419-424. DOI:10.3201/eid1703.101510.

9. Falkinham, J. O., George, K. L., Parker, B. C., & Gruft, H. (1984). In vitro susceptibility of human and environmental isolates of Mycobacterium avium, M. intracellulare, and M. scrofulaceum to heavy-metal salts and oxyanions. *Antimicrobial Agents and Chemotherapy,*25(1), 137-139. DOI:10.1128/aac.25.1.137

10. Ho, Y. S., Adroub, S. A., Aleisa, F., Mahmood, H., Othoum, G., Rashid, F., … Abdallah, A. M. (2012). Complete genome sequence of *Mycobacterium fortuitum* subsp. fortuitum type strain DSM46621. *Journal of Bacteriology*, 194(22), 6337–6338. DOI:10.1128/JB.01461-12

11. Honda J. R., Hasan N. A., Davidson R. M., Williams M. D., Epperson L. E., Reynolds P. R., … Strong, M. (2016) Environmental nontuberculous mycobacteria in the Hawaiian Islands. *Public Library of Science Neglected Tropical Diseases* 10(10): e0005068.DOI:10.1371/journal. pntd.0005068

12. Kallimanis, A., Karabika, E., Mavromatis, K., Lapidus, A., Labutti, K. M., Liolios, K., . . . Drainas, C. (2011). Complete genome sequence of *Mycobacterium sp.* strain (Spyr1) and reclassification to *Mycobacterium gilvum* Spyr1. *Standards in Genomic Sciences,*5(1), 144-153. DOI:10.4056/sigs.2265047

13. Mirsaeidi, M., Farshidpour, M., Allen, M. B., Ebrahimi, G., & Falkinham, J. O. (2014). Highlight on advances in nontuberculous mycobacterial disease in North America. *BioMed Research International, 2014*, 1–10. DOI:10.1155/2014/919474

14. Mycobacteria in the environment. *Clinics in Chest Medicine*, 23(3), 529-551. DOI:10.1016/s0272-5231(02)00014-x

15. *Mycobacterium fortuitum*. (2011). National Institute of Health. Retrieved 2017, April 15. [Web]

16. Oram, B. B. (2014). Nitrates and nitrites in drinking water and surfacewaters. *Water Research Center*. Retrieved 2017, April 15. [Web]

17. Petrini, B. (2006). Non-tuberculous mycobacterial infections. *Scandinavian Journal of Infectious Diseases*, 38(4), 246–255. DOI:10.1080/00365540500444652

18. Primm, T. P., Lucero, C. A., & Falkinham, J. O. (2004). Health impacts of environmental mycobacteria. *Clinical Microbiology Reviews*, 17(1), 98–106. DOI:10.1128/cmr.17.1.98-106.2004

19. Schinsky, M. F., Mcneil, M. M., Whitney, A. M., Steigerwalt, A. G., Lasker, B. A., Floyd, M. M., . . . Brown, J. M. (2000). *Mycobacterium septicum* sp. nov., a new rapidly growing species associated with catheter-related bacteraemia. *International Journal Of Systematic And Evolutionary Microbiology,*50(2), 575-581. DOI:10.1099/00207713-50-2-575

20. Secondary Drinking Water Standards: Guidance for Nuisance Chemicals. (2017*). Environmental Protection Agency*. Retrieved 2017, April 15. [Web]

21. Whittington, R. J., Marsh, I. B., Saunders, V., Grant, I. R., Juste, R., Sevilla, I. A., … Whitlock, R. H. (2011). Culture phenotypes of genomically and geographically diverse *Mycobacterium avium* subsp. paratuberculosis isolates from different hosts. *Journal of Clinical Microbiology*, 49(5), 1822–1830. DOI:10.1128/JCM.00210-11

22. Zemke, A. C., Shiva, S., Burn, J. L., Moskowitz, S. M., Pilewski, J. M., Gladwin, M. T., & Bomberger, J. M. (2014). Nitrite modulates bacterial antibiotic susceptibility and biofilm formation in association with airway epithelial cells. *Free Radical Biology & Medicine*, 77, 307–316. DOI:10.1016/j.freeradbiomed.2014.08.011

ABOUT THE AUTHORS

Pictured: From left to right, Olivia Voss, Dr. Jennifer Honda, and Sahana Narayan.

We are extremely grateful to the Rock Canyon Biotechnology program for offering us the opportunity to conduct college level research. From countless emails between here and Hawai'i, to driving up narrow paths and running down rocky terrain, walking on ice, and learning and doing the DNA extraction protocol enough to have it down to under 90 minutes, the skills we have learned are invaluable to our future endeavors. Be it our struggles in designing and conducting this experiment or improving our problem-solving skills with plans B through Z, and hours rewriting, this process has taught us perseverance in the face of adversity. The experience of a highly individualized and independent workspace has matured us as young adults, and our newfound sense of independence and responsibility is crucial for our success as we move forward and make the transition from a high school environment to a college setting.

In terms of future plans, Sahana aspires to go into Biology as an undergraduate before eventually going to med school and becoming a pediatric surgeon. Though her path is more clinical, she has grown to love research, and hopes to continue conducting some when obtaining her undergraduate degree. Olivia, though not directly inclined to pursue a scientific career, hopes to continue research during her undergraduate and graduate years, and plans to become involved in international affairs.

Circadian rhythm disruption effects on adipose tissue levels in *Drosophila melanogaster*

M. M. Nam, T. N. Fisher, and S. L. Fordham
Department of Science, Principles of Experimental Design in Biotechnology, Rock Canyon High School, Highlands Ranch, Colorado, USA

Obesity is currently on the rise throughout many countries in the world and studies have shown multiple factors, including genetics and environment, are contributing to this pandemic. One of these factors, a disruption in circadian rhythms, has been correlated with an increase in obesity in humans.[10] In this investigation, we tested how the disruption of the circadian rhythms due to varying amounts of light exposure throughout the entire life cycle of *Drosophila melanogaster* would impact the amount of adipose tissue, body fat, in the offspring of the *Drosophila*. We hypothesized that if F_2 larvae were exposed to constant light, they would have statistically higher amounts of body fat than larvae exposed to normal light conditions. We measured body fat by floating the F_1 larvae in a 5-10% sucrose gradient and we correlated a higher percentage of larvae floating with a higher amount of body fat in the F_2 larvae. While most of the larvae sunk at the 5% sucrose concentration in each treatment, we saw a significant difference in the percentage of larvae that floated in the 10% sucrose solution. In the constant dark treatment, 99% of the larvae floated as compared to the 80% of larvae that floated in the control. In the constant light treatment, 97% of the larvae floated as compared to the 80% of larvae that floated in the control. We found that there was a significant difference between the control and both treatments at the 10% sucrose solution but not at the 5% sucrose solution. Since there was a significant difference between both treatments compared to the control, we conclude that there is a correlation between the disruption of circadian rhythms and an increase in the relative amount of body fat in *Drosophila*. Our research strengthens the correlation between a disrupted circadian rhythm and an increase in amount of body fat.

Obesity is a disorder described as the overaccumulation of body fat, also known as adipose tissue, which is caused by a number of different factors.[10] Alarms have been raised in the medical field concerning the increasing rates of obesity found in the populations of adults and children within the United States. Consequently, many researchers are seeking methods of prevention and possible treatments. Between the years 2011 and 2014, the National Health and Nutrition Examination Survey classified approximately 17% of children in the United States, aged 2 to 19, as obese.[16] Obesity is a substantial health concern, as it leads to further health complications such as coronary heart disease, high blood pressure, type 2 diabetes, sleep apnea, and reproductive issues.[20]

According to the Center for Disease Control and Prevention, more than 50 different genes are associated with obesity, making it a challenge to understand and treat.[4] However, the condition is also influenced by environmental effects, such as sleep disturbance and irregularity of circadian rhythms.

Located in the suprachiasmatic nuclei of the anterior hypothalamus of the brain in mammals, our circadian rhythm is often referred to as the clockwork in which the body functions (**Fig. 1**). Light is detected from the retina's photosensitive retinal ganglion cells (RGCs), which in turn respond to light without the need for the rods and cones. The RGCs spread their axons to many regions of the brain. From there, the information travels through the retino-hypothalamic tract, and then stops at the suprachiasmatic nuclei.[7] The body's circadian rhythm regulates metabolism and activates metabolic enzymes, maintains body temperature, and releases hormones. The rhythmic nature of several metabolic hormones such as glucagon, ghrelin, leptin, and insulin may also exist.[13]

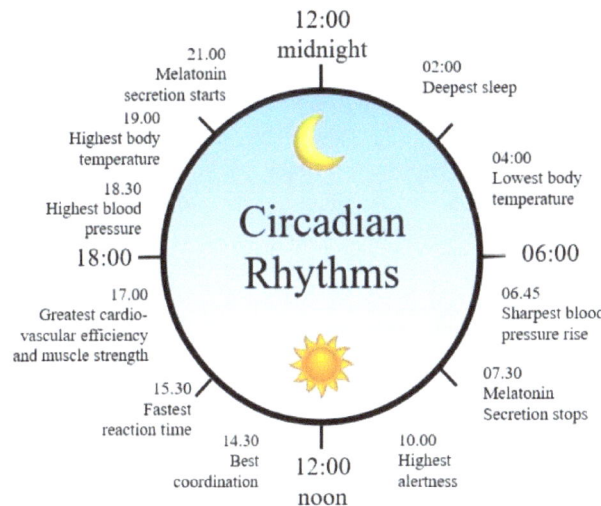

Figure 1: An average circadian rhythm for humans during a 24-hour period.[2]

A disruption of any kind has a great possibility of causing a homeostatic imbalance in the body. Evidence of this imbalance can be seen in employees who work night shifts, truck drivers, and frequent travelers suffering from jet-lag. These groups have an increased risk of developing serious medical conditions such as obesity, coronary disease, and cancer.[13] Certain stimuli, such as artificial light from electronics, can also disrupt the regularity of the mammalian circadian rhythm.[10]

Many key hormones play a role in the metabolic processes of humans. One hormone, leptin, is crucial when studying obesity. Leptin suppresses hunger and works in tandem with ghrelin, a hormone which initiates the feeling of hunger (**Fig. 2**). If leptin levels are altered and not in balance with its counterpart, increased appetite may lead to obesity.[1]

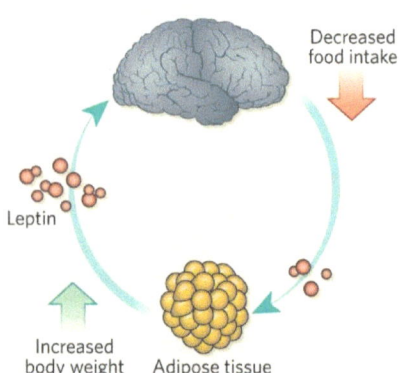

Figure 2: The mechanism of leptin in the conditions of weight loss or weight gain.[9]

Although leptin is thought to be found solely in vertebrate species, researchers at the Harvard Medical School uncovered a molecule very similar to leptin in their *Drosophila* models. They found that the molecule was a specific protein, named Upd2. When knocking out this protein, they found that the flies behaved as if they were starving, even though they had eaten a typical amount of caloric content.[3] Through their work, the researchers concluded that Upd2 is a close homolog to leptin in *Drosophila*. With this information, we believe that our research with *Drosophila* is also applicable in humans, giving us the means to research this topic, which is increasingly becoming more prevalent in our society.

Another equally important hormone in the discussion of the human circadian rhythm is melatonin (**Fig. 3**). This hormone is found in humans and *Drosophila*. It is secreted by the pineal gland in the brain and regulates our body's biological clock by keeping it synchronized, as well as managing blood pressure. It also responds to different light environments in a unique way. For instance, melatonin levels increase in darkness, creating the feeling of drowsiness, while the levels decrease in more light, creating a sense of alertness. Exposure to bright light in the typical hours of darkness can alter the body's circadian rhythms.[8] Electronics, including a computer, and television screens, in the late hours of the night are thought to alter sleep patterns. 72% of young adults aged 13 to 18 use their electronic devices immediately before they go to bed, and many teenagers report frequent sleep disturbances which keep them from feeling restful and rejuvenated the next day.[11] The lack of sleep can greatly inhibit one's productivity and mood. Studies have also shown that the presence of melatonin is necessary for the production and secretion of

insulin, a very crucial hormone that plays a role in the regulation of glucose levels in blood.[5] This metabolic hormone is also the main factor in diabetes, a disease that can be found in approximately 208,000 Americans.[18]

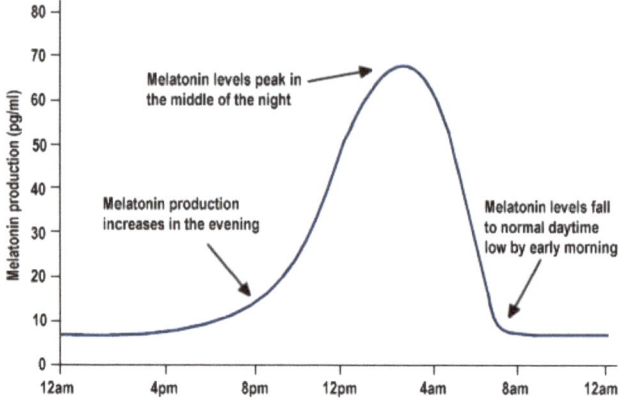

Figure 3: The amount of melatonin secreted at certain times of the night.[14]

Our research tested the impact of disruption of circadian rhythm due to irregular light exposure on the relative amount of body fat observed in the offspring of *Drosophila*. We used wild type, Oregon R strain of *Drosophila melanogaster* (fruit flies) for this experiment (**Pic. 1**). *Drosophila* is an excellent model organism to use in this research because they reveal a stronger phenotype and have a less repetitive genome than humans or mice.[6] Furthermore, *Drosophila* contain approximately 77% of genes in common with humans, including the

Picture 1: Wild Type, Oregon R strain *Drosophila melanogaster.*[12]

adipose gene that is important in the metabolism of fruit flies, mice, and humans.[13,19] Fruit flies also have the Upd2 protein, which has been shown to serve as a homolog to leptin in humans.[3] In prior research, *Drosophila* was used to research metabolic homeostasis, obesity related diseases, and diabetes.[19] In our research, we exposed *Drosophila* to a regular light cycle (comprised of 13 hours light followed by 11 hours dark), constant light, and constant dark conditions throughout their entire life. We then measured the body fat relative to the controls in the offspring using a sucrose gradient and we correlated a higher percentage of larvae floating with a higher amount of body fat in the F_2 larvae. We were able to draw this correlation due to past research done by Jeremy Mosher and our verification trial. For any sucrose concentration from 8-12%, larvae with a higher amount of body fat floated in greater numbers than normal larvae.[15] This trend was also observed in our verification trial with our genetically fat larvae floating in greater numbers than our control larvae.

METHODS

In this experiment, we tested how the disruption of circadian rhythms affected the floatation of F2 larvae in a sucrose gradient as a measure of body fat in *Drosophila melanogaster*. In each trial, three different vials were set up in each of the three treatment areas which were constant darkness, constant light, and a 13 hour light:11 hour dark cycle. This last treatment best mimicked natural light conditions and served as our control.

Experimental Design

We created our three treatment areas on a light cart originally used for plants and kept the cart in the RCHS laboratory prep room at a fairly constant temperature of approximately 25°C. We built the constant dark treatment area by wrapping a cardboard box with a black cloth and placing it into a plastic bin. To ensure that no light could enter from beneath the box, we taped a textbook on top of the box to weigh it down. We set this bin on the top most shelf of the plant cart (**Pic. 2**). On the shelf directly below, we set up our constant light treatment area which remained lit 24 hours a day. For our control, 13h light:11h dark, environment, we used the bottom shelf on the plant light cart and placed the vials in the center of the lamp to minimize light exposure from the constant light treatment areas. We also placed a white platform between the two shelving areas to minimize light disruption to the other treatment on the bottom shelf. We housed all of the treatment areas on the plant light cart to keep the variables, such as temperature and humidity, constant (**Pic. 2**).

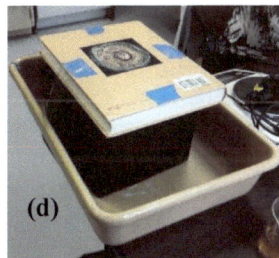

Picture 2: The plant cart with the constant dark (**a**), constant light (**b**), and control (**c**). Also a close up at the constant dark treatment area (**d**).

All of the culture vials were prepared using 4g/mL Formula 4-24® Instant *Drosophila* Medium from Carolina Biological with approxi-mately 5-6 grains of yeast added to the surface. Approximately 20 wild type *Drosophila* obtained from Carolina Biological were transferred into each vial to create the first generation of our research (**Pic. 3**). Nine vials in total were prepared and three vials were then immediately placed in each of the three different treatment areas. After three days, we released all of the adult flies and placed the vials back into their respective treatment areas. We kept these vials in these conditions for 14 days in order to allow for a complete life cycle to take place in the vials. The F1 flies that had hatched from this starter generation had been exposed to the treatment throughout their entire life cycle (from egg through the adult stage).

After the F_1 generation hatched, we transferred them to six new vials, with approximately 20 flies each, which we then placed back in their original treatment areas. After three days, we released all the adult flies and placed the vials back into their treatment areas. We waited another 10 days to allow the F_2 offspring to develop into larvae. We selected three vials from each treatment area and performed a density assay on the larvae in that vial. After we collected data from the three vials, we determined the average and recorded it as one trial. We repeated this process three more times for a total of four trials.

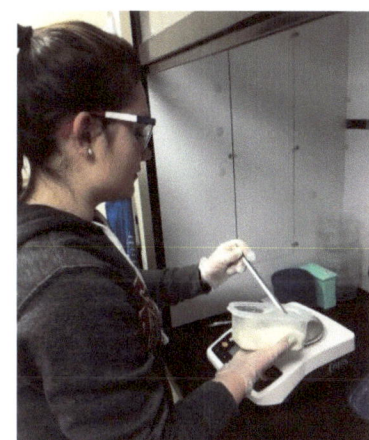

Picture 3: Fisher preparing culture vials by making *Drosophila* media for each vial.

We correlated a higher percentage of larvae floating with a higher amount of body fat in the F_2 larvae and in order to validate this correlation and our density assay, we created an obese phenotype model of *Drosophila*. To do this, we crossed virgin female *Drosophila* with the UAS iSpen RNAi genotype with the males having the dcg>Gal4 genotype creating dcg>iSpen F_1 offspring with an obese phenotype. This phenotype had previously been shown by the Reis Lab to exhibit a high amount of body fat in comparison to the wild type.[17] To serve as our control, we also crossed virgin females with a UAS iw RNAi genotype with males having the dcg>Gal4 genotype to create F_1 dcg>iw control pheno-type offspring. These flies still had the dcg>Gal4 genotype, but they did not have the high body fat that is observed in the previous group. Once the F_1 larvae began to wander, the F_1 larvae were floated using the same sucrose gradient protocols as in our experimental trials.

Picture 4: Larvae floating at equilibrium in a 20% sucrose solution.

Density Assay

To compare the relative amounts of fat in the F_2 larvae, we used a sucrose density assay. We followed the protocols obtained from the Reis Laboratory of the University of Colorado Denver Anschutz Medical Campus, but slightly adjusted their protocols.[17] We added 40 mL of a 20% sucrose solution, which was created with 1X PBS, into the culture vials containing the wandering larvae and waited approximately 20 minutes to allow equilibrium to be established **(Pic. 4)**. Once all of the larvae had floated to the top, we extracted them from the vial using forceps and transferred them into a 10% sucrose solution. We then recorded the number of larvae that were floating **(Pic. 5)**. Next, we lowered the density to a 5% sucrose solution and again recorded the number of larvae floating. We performed the density assay for our experimental trials and our validation trial.

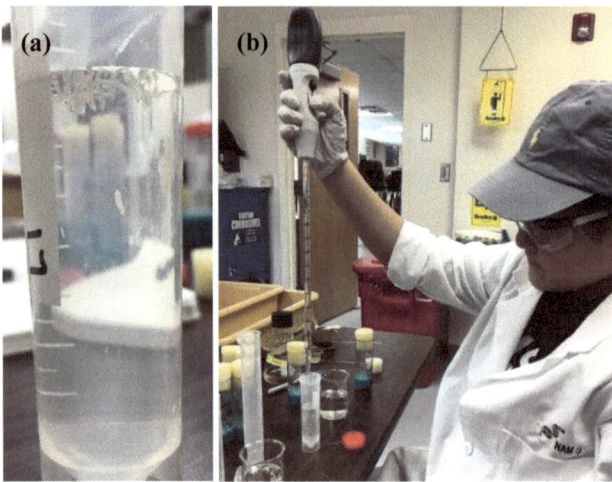

Picture 5: (a) Nam pipetting a 10% sucrose solution into **(b)** a conical tube with F_2 larvae.

RESULTS

In order to determine a correlation between a disrupted circadian rhythm and the relative amount of body fat in the *Drosophila* larvae, we counted the number of F_2 larvae floating in 5% and 10% sucrose solutions. Using this data, we compared the light and dark treatments to the control and ran a z-test of proportions to identify any significant differences between the control and two treatment areas.

For the light treatment, 97% of the larvae floated at a 10% concentration but only 80% of the larvae floated in the control. At the 5% concentration, we found that there was no difference between the control and light treatment **(Graph 1)**. To verify this difference at the 10% concentration, we ran a z-test of proportions and found that the light treatment was statistically higher than the control based on the p-value of 0.9959.

Similar to the constant light treatment, the constant dark treatment resulted in 99% of the larvae floating in the 10% concentration and there was no difference observed between the control and the constant dark treatment at the 5% concentration **(Graph 2)**. When we ran the z-test, we found that the dark treatment was also statistically higher than the control value, with a p-value of 0.9988, at the 10% concentration.

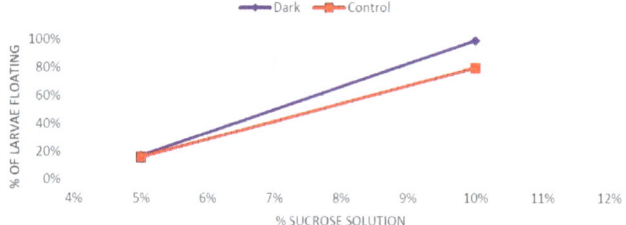

Graph 2: The average percentage of larvae that floated in a 5-10% sucrose gradient for the dark and control trials.

When comparing our light and dark treatment data to our validation test using known phenotypically fat (iSpen>dcg) and normal/control (iw>dcg) *Drosophila*, we found that they also show a similar increase in the percent of larvae that float at the 10% sucrose concentration. For the fat phenotype, 97% of the larvae floated at the 10% concentration and only 64% of the larvae floated in the control. However, unlike our data, these larvae also showed a difference at the 5% concentration. For the fat phenotype, 44% of the larvae floated at the 5% concentration and 14% of the larvae floated in the control. The fat phenotype was able to reach higher percentages of larvae floating in lower sucrose concentrations than the control line which indicates a more significant difference in body fat than our experimental results **(Graph 3)**. When we ran the z-test, for this validation test and we found that the fat phenotype was significantly higher than the control at both the 5% and 10% concentrations with p-values of 0.997 and 1.0 respectively.

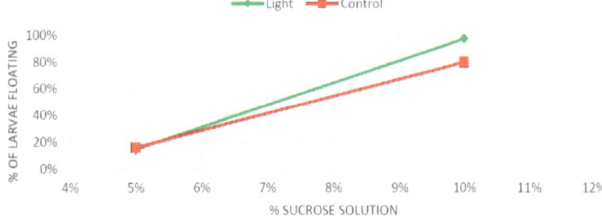

Graph 1: The average percentage of larvae in that floated in a 5-10% sucrose gradient for the light and control trials

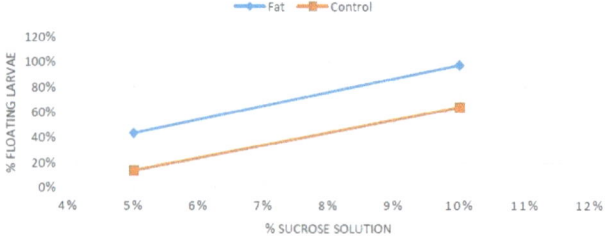

Graph 3: The percentage of larvae that floated in a 0-20% sucrose gradient for the verification trial. The fat is the iSpen>dcg genotype and the control is the iw>dcg genotype.

DISCUSSION

Obesity is a dangerous disorder and its prevalence is on the rise due to many factors such as genetics and the environment. In this experiment, we tested the effects of disrupting one of these factors, circadian rhythm, on the relative amount of body fat in *Drosophila melanogaster*. We hypothesized that increased exposure to light, as compared to a normal cycle, would result in an increase in the relative amount of body fat in *Drosophila*. We found the relative body fat by floating the F_2 larvae in a 5% and 10% sucrose solution and correlated a higher percentage of larvae floating with a higher amount of body fat in the F_2 larvae.

During our trials, we found that there was a significant increase in larvae that floated between the control and both treatments in the 10% sucrose solution but none of the treatments had a significant difference with the control in the 5% sucrose solution. In the light treatment 97% of the larvae floated at the 10% concentration, which is statistically different from the 80% of the larvae that floated in the control. In the dark treatment, 99% of the larvae floated in the 10% concentration, which is statistically different from the 80% of the larvae that floated in the control. While both our verification and experimental trials showed a significant difference at the 10% concentration, only the validation trial showed a difference at the 5% concentration. This could indicate that the larvae were not fat enough to float at the 5% concentration.

While considering our results, statistical analysis, and errors, we are able to conclude that there is a correlation between the disruption of circadian rhythms and an increased amount of body fat in *Drosophila*. Our data showed that there is a significant difference in the amount of fat between F_2 larvae that have been exposed to constant darkness or constant throughout their life span when compared to F_2 larvae under normal light conditions. In previous research, scientists tested to see which brain regions were necessary to maintain body fat in *Drosophila*. They found that they could induce high amounts of body fat in larvae when they deactivated the neurons in specific brain regions. They found that the larvae had a higher amount of body fat as measured by the same density assay we used. In the 10% sucrose solution, approximately 70% of the "fat" larvae floated, whereas only approximately 40% of their control larvae floated, compared to 99% and 80% respectively of our larvae at the same concentration.[15] While their experiment resulted in a greater difference in the larvae from the control group, our results still show a statistically significant difference.

The fact that our data did not exhibit any significant difference at a 5% sucrose solution was unexpected because in the validation trials, there was a significant difference the number of larvae floating in both the 5% and 10% sucrose solutions. This might indicate an error in our experimental protocols or that our experimental larvae were not fat enough to float at the 5% sucrose concentration. We accept our hypothesis in that the light treatment caused an increase in body fat but we did not expect the dark treatment to have the same effect. According to our data, instead of only a constant exposure to light, a constant exposure to darkness can also cause an increase in the relative body fat for F_2 larvae.

A potential source of error in our experiment was the dry condition of the media in our control treatment area vials for one of the trials. This could have altered our data because there were not as many larvae in these vials that could be used to collect data. However, this error had a minimal effect because we still had data from the three other trials. Another source of error was that the control area was underneath the constant light treatment area. The light from the top shelf illuminated the entire room to a small extent. This could have impacted our control during the night period and we may have observed an even greater difference between our treatments and control without this variable. Another source of error was that we used water instead of PBS when diluting the 10% sucrose solution to a 5% sucrose solution which caused the actual sucrose concentration to be a bit lower than 5%. This would cause more larvae to sink than in an actual 5% sucrose solution. Without this error we may have seen a difference between the treatments and the control at the 5% concentration.

The next steps for this research would be to repeat this experiment again to identify how a more distinct separation of the treatment areas would affect the data. A repeat of this experiment, with these changes, is necessary for this research to further progress, because our conclusion should be validated so that our results can be seen as credible. Then, the research should be validated with mammals, such as mice, so that our research can be more applicable to humans.

ACKNOWLEDGMENTS

We would like to thank Stephanie Bonney and Santiago Fregoso of the Stem Cells and Development Graduate Program at the University of Colorado, Denver for their mentorship and unwavering support throughout this research. We would also like to thank Kelsey Hazegh and Cayla Jewett, also from CU Denver, for providing us with the genetically obese *Drosophila* to use as models for obesity during our assay validation trial. This experiment could not have taken place without the generous donation from Julie Fisher, and we are very grateful. We would like to recognize Bryan Winkelman for his expansive technical support, including the production of the research website, his guidance with blog posts, publishing our research in a class journal, and helping organize our ability to fundraise. We would like to acknowledge Jim McClurg and Amy England for producing our video to introduce our research, and Dr. Jason Dunkle for providing advice when determining the statistical analysis of our data. We thank Wendy Lerolland for her editorial assistance during the writing of this article. We would also like to thank Tom Dillon with helping us develop our research and giving us feedback and advice on our experimental design. We also appreciate Suzanne Petri and Nikki Dobos for allowing us to share their lab space. Lastly, we thank Rock Canyon High School for providing laboratory space and the equipment used in this experiment.

REFERENCES

1. Ahima, R. S. (2008). Revisiting leptin's role in obesity and weight loss. *Journal of Clinical Investigation*. DOI:10.1172/jci36284

2. Bullock, J. (2016). Two-minute explainer: Circadian rhythms. Lux Review. Retrieved on 2017, March 15. [Web]

3. Cameron, D. (2012). Obesity-Related Hormone Discovered in Fruit Flies. Harvard Medical School. Retrieved on 2016, September 25. [Web]

4. Choquet, H., & Meyre, D. (2011). Genetics of Obesity: What Have We Learned? *Current Genomics*, 12(3), 169–179. DOI:10.2174/1389202117 95677895

5. Cipolla-Neto, J., Amaral, F. G., Afeche, S. C., Tan, D. X., & Reiter, R. J. (2014). Melatonin, energy metabolism, and obesity: a review. *Journal of Pineal Research*, 56(4), 371-381. DOI: 10.1111/jpi.12137

6. Cirelli, C. (2009). The genetic and molecular regulation of sleep: from fruit flies to humans. *Nature Reviews Neuroscience*, 10(8), 549-560. DOI: 10.1038/nrn2683

7. Do, M. T. H., & Yau, K. W. (2010). Intrinsically Photosensitive Retinal Ganglion Cells. *Physiological Reviews*, 90(4), 1547-1581. DOI:10.1152/ physrev.00013.2010

8. Fonken, L. K., & Nelson, R. J. (2014). The Effects of Light at Night on Circadian Clocks and Metabolism. *Endocrine Reviews*, 35(4), 648-670. DOI: 10.1177/0748730413493862

9. Friedman, J. M. (2010). A Tale of Two Hormones. Nature Medicine. Retrieved on 2017, March 31. [Web]

10. Froy, O. (2010). Metabolism and circadian rhythms—implications for obesity. *Endocrine Reviews*, 31(1), 1-24. DOI: 10.1210/er.2009-0014

11. Infographic: How Technology Affects Our Sleep. (2013). Rasmussen College. Retrieved on 2016, October 24. [Web]

12. Jennings, B. H. (2011). *Drosophila*–a versatile model in biology & medicine. *Materials Today*, 14(5), 190-195. DOI: 10.1016/S1369-7021 (11)70113-4

13. Lopez-Minguez, J., Gómez-Abellán, P., & Garaulet, M. (2016). Circadian rhythms, food timing and obesity. *Proceedings of the Nutrition Society*, 1-11. DOI: 10.1017/S0029665116000628

14. Mastin, L. (2013). How Sleep Works. Sleep. Retrieved on 2017, March 15. [Web]

15. Mosher, J., Zhang, W., Blumhagen, R. Z., D'Alessandro, A., Nemkov, T., Hansen, K. C., . . . Reis, T. (2015). Coordination between Drosophila Arc1 and a specific population of brain neurons regulates organismal fat. *Developmental Biology*,405(2), 280-290. DOI:10.1016/j.ydbio.2015.07.021

16. Patel, S. R., & Hu, F. B. (2008). Short sleep duration and weight gain: a systematic review. *Obesity*, 16(3), 643-653. DOI:10.1038/ob y.2007. 118

17. Reis, T., Van Gilst, M. R., & Hariharan, I. K. (2010). A buoyancy-based screen of *Drosophila* larvae for fat-storage mutants reveals a role for Sir2 in coupling fat storage to nutrient availability. *PLoS Genet*, 6(11), e1001206. DOI: 10.1371/journal.pgen.1001206

18. Statistics About Diabetes. (2016). American Diabetes Association. Retrieved on 2016, October 24. [Web]

19. Tompkins, L., Cardosa, M. J., White, F. V., & Sanders, T. G. (1979). Isolation and analysis of chemosensory behavior mutants in *Drosophila melanogaster*. *Proceedings of the National Academy of Sciences*, 76(2), 884-887. National Institutes of Health. Retrieved on 2016, September 22. [Web]

20. Xu, K., Zheng, X., & Sehgal, A. (2008). Regulation of Feeding and Metabolism by Neuronal and Peripheral Clocks in *Drosophila*. *Cell Metabolism*, 8(4), 289-300. DOI: 10.1016/j.cmet.2008.09.006

ABOUT THE AUTHORS

Pictured: From left to right, Taylor Fisher, Michelle Nam with Brenna Clay and mentors Santiago Fregoso and Stephanie Bonney.

Over the course of our research, we believe that we have grown academically, as well as personally. The amount of independent work this course requires has taught us to be successful when working in a partnership, as well as individually to meet a common goal. We felt a great sense of humility when our ideas were being critiqued, and learned how to use such constructive criticism to improve what was lacking, whether it was an area of our experiment or ourselves.

Although we worked as peers, we both have gained crucial skills on how to lead on areas in which we feel particularly confident. Because of this, we were able to proficiently perform what was asked of us and we were content with our results.

We also gained valuable skills from working in this type of academic setting. The protocols needed for our research have taught us how to operate on a level similar to that of a professional laboratory. Also, our goal to create a novel idea for research helped us refine our critical thinking skills, and communicate with others currently working in the professional scientific field. Working with talented graduate students, such as Ms. Bonney and Mr. Fregoso has been an amazing experience, and we can say that without a doubt we have been inspired to continue our studies in the field of science. With the knowledge we have obtained throughout this experience, we feel very prepared on what we have in store for the future.

The effects of green tea extracts on the trehalose levels in a diabetic model of *Drosophila melanogaster*

J. R. Olcott, S. A. Nasseth, and S. L. Fordham
Department of Science, Principles of Experimental Design in Biotechnology, Rock Canyon High School, Highlands Ranch, Colorado, USA

Type 1 diabetics are unable to produce insulin, thus sugar cannot enter cells and remains in the blood, causing hyperglycemia. Untreated diabetics exhibit elevated levels of circulating sugar. Green tea is widely consumed across Eastern and Western cultures for its health benefits. Previous research has shown that the main polyphenol in green tea, epigallocatechin gallate (EGCG) has anti-hyperglycemic effects.[2] Our study tested the effects of adding green tea extract with a high concentration of EGCG to the food of a type 1 diabetic model of *Drosophila melanogaster*. After creating diabetic *Drosophila*, we measured the trehalose levels in the wild type *Drosophila* and compared it to the diabetic fly model. No statistical difference was found between the trehalose levels of the diabetic flies in the green tea extract and media only. The trehalose level of wild type flies in green tea extract, however, was significantly lower than the trehalose level of the wild type flies in media only. This data shows that the green tea extract may have a natural decreasing effect on the trehalose level in wild type *Drosophila*, but does not seem to be helpful in the diabetic model. Due to the lowered blood sugar levels of wild type flies, our data supports the idea that green tea extracts may be able to help prevent the development of type 2 diabetes by maintaining an overall lower blood sugar level in the blood of nondiabetics.

29.1 million Americans have a type of diabetes (**Pic. 1**). Approximately 1.25 million Americans have type 1 diabetes.[7] Every year, approximately 18,000 American youth are diagnosed with type 1 diabetes.[12] Currently, the treatment for this disease is limited and diabetics must manage their blood sugar by injecting themselves with insulin up to ten times a day, with many becoming resistant over time due to high usage.

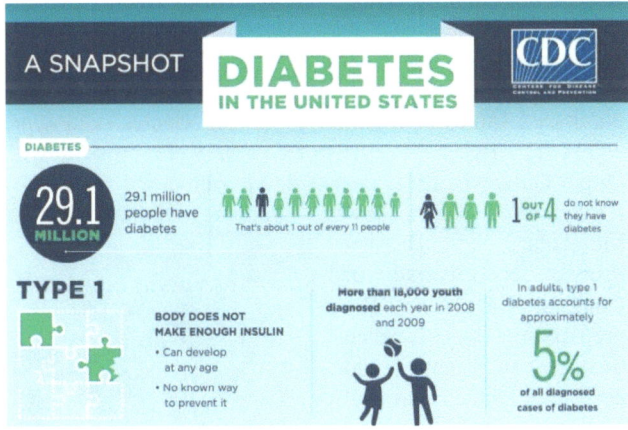

Picture 1: Pictured above are statistics of diabetes in the United States, provided by the Centers for Disease Control and Prevention. 29,100,000 Americans have diabetes, which is about 1 in every 11 Americans. There are approximately 18,000 Americans diagnosed with type 1 diabetes each year.[12]

The pancreas regulates levels of sugar in the bloodstream by secreting insulin and glucagon. Insulin is an enzyme that binds to receptor cells in all cell membranes and allows sugar to be transported into the cell for use in energy production. Without insulin, digested sugars stay in the bloodstream. The beta cells in the islets of the pancreas of type 1 diabetics fail to produce insulin, and it is thought that in most cases, the immune system destroys these cells in the pancreas.[3] This immune response may be due to genetics and/or environmental factors. Without insulin, sugar accumulates in the blood, resulting in potential systemic damage to the eyes, kidneys, blood vessels, and nerves. People with type 1 diabetes must regularly inject themselves with insulin in order to survive.[6]

Research is being conducted to identify alternate treatments for diabetes, such as the use of stem cell therapy. In stem cell therapy, scientists convert adult stem cells from diabetic patients into pancreatic beta cells. These cells then produce insulin in response to elevated levels of sugar in the bloodstream. The self-renewing capacity of human pluripotent stem cells and their ability to make insulin-producing beta cells has encouraged many researchers to make glucose-responsive beta cells in the laboratory.[8] However, scientists are currently unable to transplant the stem cells into the pancreas without triggering an immune response that would again kill all the insulin producing beta cells.[13]

Although abundant research has been conducted on the effects of green tea extracts on type 2 diabetic models, very little has been researched on type 1 diabetes. One study found that patients with hyperglycemic-related hypertension, given a supplemented diet of green tea with EGCG for three months, showed results nearly as effective as treatment with Enalapril, a medication used to treat high blood pressure, diabetic kidney disease, and heart failure.[16]

In this experiment we used *Drosophila melanogaster* flies modified to mimic type 1 diabetes to assess how a diet high in green tea extract would affect circulating blood sugar levels in adult flies. The insulin-producing cells in the *Drosophila* were destroyed by apoptosis. *Drosophila* are good diabetic models because their metabolism parallels the structure and function of humans. Humans and flies have homologous organ systems: the gut absorbs nutrients and body fat stores nutrients and functions as a nutrient sensor. Even though *Drosophila* have an open circulatory system whereas humans have a vascular blood system, the *Drosophila* heart is essential for the circulation of nutrients.[1] Flies express eight *Drosophila* insulin-like peptides (Dilp), which are proteins with high sequence similarity to insulin. Because of this, we were able to study the effects of green tea extracts on the blood sugar of diabetics as it pertains to human physiology.[9]

A study conducted on the metabolism of *Drosophila* compared the circulating trehalose and glucose levels of larvae and adults. In humans, the circulating blood sugar is glucose, whereas in *Drosophila,* the more prominent circulating blood sugar is trehalose. In flies, glucose was present in much smaller quantities. The normal trehalose levels in both larvae and adults were found to be 20 mg/mL, whereas the glucose levels were variable between the two life stages: less than 1 mg/mL in larvae, and 6 mg/ml in adults.[4]

		Parent 1	
		UAS Reaper	**UAS Reaper**
Parent 2	**Dilp2>Gal4**	Dilp2 Reaper *(diabetic)*	Dilp2 Reaper *(diabetic)*
	Curly-O	UAS Reaper CyO *(curly winged & not diabetic)*	UAS Reaper CyO *(curly winged & not diabetic)*

Table 1: This table represents a genetic cross of the flies we used. We crossed a *Drosophila* with the Dilp2 gene on one chromosome and the Curly-O gene on the other with a homozygous reaper. The progeny were prominently diabetic flies. The Dilp2 x Reaper flies were diabetic. The diabetic flies were represented by straight winged flies, and the non-diabetic were represented by the curly winged flies.

In order to create the diabetic *Drosophila* model, we crossed homozygous *Drosophila* with the reaper gene, which increases the rate of apoptosis, with heterozygous *Drosophila* with the Dilp2 gene and a Curly-O gene, which is a curly wing gene. When *Drosophila* have both the reaper gene and the Dilp2 gene, apoptosis is increased in the insulin-producing cells, making the flies diabetic.[11] When the Dilp2 flies are crossed with the reaper flies, half of the offspring have the Dilp2 and reaper genes, making them diabetic. The other half of the offspring should have the Curly-O and reaper gene, making them non-diabetic curly winged flies **(Table 1)**. When the *Drosophila* are homozygous for the Dilp2 gene, they become sick and do not reproduce normally. We crossed the Dilp2 gene with

the Curly-O gene so that the non-diabetic *Drosophila* were distinguishable from the diabetic models. The offspring that are diabetic have normal wings, and the non-diabetic flies will have curly wings.

After exposing the *Drosophila* to the ECGC by adding green tea extracts to the food, we compared the trehalose levels of the diabetic flies to the wild type flies with a Hexokinase (HK) assay kit. If the green tea exhibits anti-hyperglycemic effects in diabetic flies, then green tea supplements could be used diabetics to manage their blood sugar levels easier.

METHODS
Pre-Trials
For our pre-trials, we prepared vials of Carolina Biological instant fly media and varying concentrations of green tea extracts. We used a standardized green tea extract product from TeaVigo containing 94% epigallocatechin gallate (EGCG). Each TeaVigo capsule contained 150 mg of green tea extract. We tested fly cultures with 1-capsule, 2-capsule,

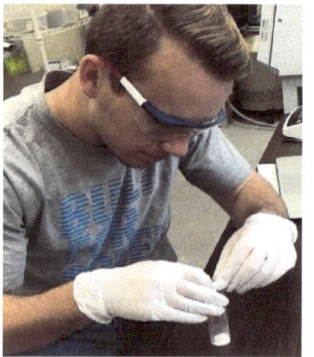

and 3-capsule concentration of TeaVigo added to 3.5g of fly media **(Pic. 2)**. Then we added 15 mL of water to each mixture. We tested each concentration with wild type flies to ensure the tea did not cause harm or reduce reproduction.

The highest concentrated vials resulted in dried media and lowered reproduction, so we decided to use the 2-capsule concentration in our experimental trials.

Picture 2: Olcott breaks open a capsule of green tea to mix with the fly media.

Experimental Trials
In our experimental trials, UAS reaper *Drosophila* flies, purchased from the Bloomington Stock Center (genotype: w[1118]; P{w[+mC]=UAS-rpr.C}14), were crossed with Dilp2>Gal4/cyo flies provided by our mentors Kelsey Hazegh and Cayla Jewett, graduate students in the Molecular Biology Program at University of Colorado Denver. We prepared two vials for each treatment so that we could perform both reciprocal crosses: control treatment (fly media only) and the experimental treatment, consisting of 150 mg green tea extract. Virgin female flies were collected from both fly lines and placed into the treatment vials **(Pic. 3)**.

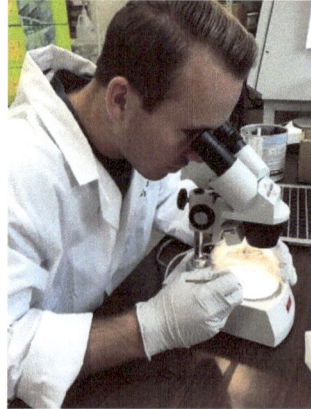

Picture 3: Olcott looks at flies under a microscope to determine their gender and type.

For each treatment, one vial contained 10-15 virgin female DILP2 flies crossed with four male Reaper flies, while the second vial contained six-eight female Reaper flies crossed with three male DILP2 flies. A vial of each treatment was also prepared with wild type flies to test the effects of each treatment on the blood sugar of wild type flies for comparison.

Once pupae were observed in each vial, the adults were released. When the F_1 flies of the Dilp2>Gal4/cyo and UAS Reaper flies hatched, we tested the trehalose levels of the male straight-winged diabetic flies. Male and female flies have differing trehalose levels, so only males were collected for testing. When the F_1 wild type flies hatched, we also tested the trehalose levels of the male flies.

Trehalose Assay

We collected 20 samples of wild type flies in the green tea extract, of wild type flies in media only, and of diabetic flies in media only. Only 8 samples were able to be collected of the diabetic flies in the green tea. Each sample contained five male flies. The flies were euthanized with dry ice and then washed with a 1x PBS solution. After removing the PBS, we added 100 uL of ice-cold trehalose buffer **(Pic. 4)** and homogenized the samples with a pestle for 2-3 minutes **(Pic. 5)**.

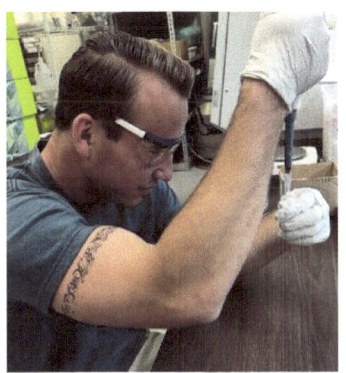

Picture 4: Olcott pipettes the trehalose buffer into the tube of flies.

After homogenization, we placed the samples in a 70ºC water bath for 10 minutes to denature all of the natural enzymes in the samples **(Pic. 6)**. Then, we centrifuged the samples at 10,000 rpm for three minutes in a 4ºC refrigerated microcentrifuge. We stored the supernatant at -20ºC until all samples were ready to be tested. Samples were tested for glucose and

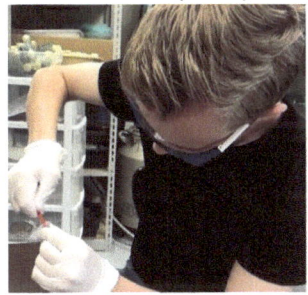

Picture 5: Nasseth homogenizes the sample by grinding with a pestle.

trehalose using the Hexokinase assay kit from Sigma Aldrich. Once we had collected 20 samples from each treatment, we prepared trehalose and glucose standard solutions of 0 mg/mL, 0.01 mg/mL, 0.02 mg/mL, 0.04 mg/mL, 0.08 mg/mL, and 0.16 mg/mL for the trehalose quantification protocol.

When the standards were made, 30 uL of the supernatant from each sample was tested for glucose and a second 30 uL tested for trehalose. 30 uL trehalose buffer was added to each glucose test sample and 30 uL of trehalose solution was added to each trehalose test sample. The trehalose solution consisted of trehalase, which breaks down the

trehalose into glucose. The glucose sample measures the glucose concentration in the flies, excluding trehalose. The standards and samples were then incubated at room temperature for 24 hours in order for the trehalase to break down the trehalose into glucose.[15]

After incubation, we centrifuged all samples and standards for three minutes and then added 1 mL of the

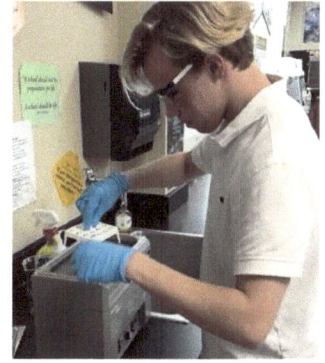

Picture 6: Nasseth places each sample into the heat bath set to denature them.

HK assay reagent to each sample and standard. We then incubated the samples and standards at room temperature for 15 minutes before measuring the standards.

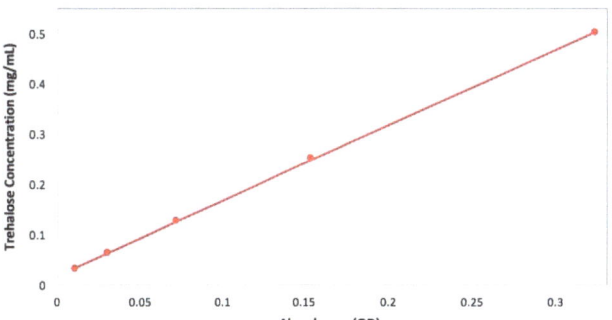

Graph 1: This graph shows the absorbance measurements of our known trehalose concentration standards. We converted the absorbance of the trehalose samples into trehalose concentrations using this graph. We also used a similar graph for converting the absorbance measurements of the glucose samples into glucose concentrations.

We quantified the trehalose levels using a nanodrop and measured the absorbance readings (OD) at 285 nm, to create a standard curve **(Graph 1)** for trehalose **(Pic. 7)**. We converted the absorbance of the trehalose sample into a trehalose concentration using the trehalose standard curve by approximating the concentration (mg/mL) based on the absorbance reading of our glucose and trehalose standards **(Graph 1)**. We then converted the glucose absorbance (OD) into a glucose concentration (mg/mL) using a similar glucose standard curve. The glucose concentration was then subtracted from the trehalose concentration to calculate the free trehalose concentration for each sample.[15]

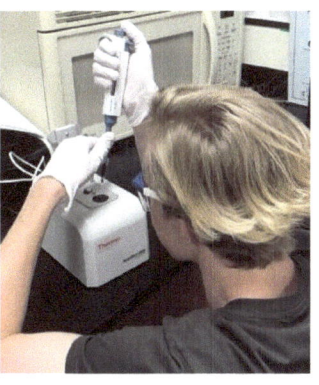

Picture 7: Nasseth pipettes the standard solution on the pedestal of the nanodrop to get a reading.

RESULTS

After preparing the samples of each group from both treatments, we measured the absorbance at 285 nm. We then subtracted the glucose concentration from the trehalose concentration to calculate the free trehalose concentration, which is the circulating sugar in *Drosophila*.

The average trehalose concentration of the wild type flies in media only was 0.1076 mg/mL, while it was much lower in the green tea extract, at 0.0781 mg/mL. The average trehalose concentration of the diabetic flies in media only was higher than observed, with wild type flies as expected at 0.1267 mg/mL. However, contrary to our hypothesis, the trehalose concentration did not change in the green tea extract and remained high at an average of 0.1287 mg/mL **(Graph 2)**.

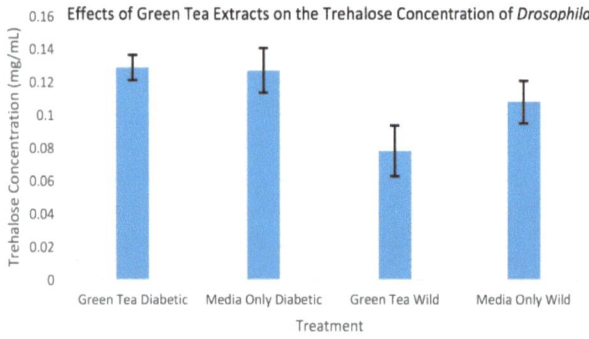

Graph 2: The graph shows the trehalose concentrations of the diabetic and wild type flies in the green tea extract and media only treatments. The graph also shows ±2SEM, which is a 95% confidence interval for each group in each treatment.

We ran a T-test of the average free trehalose concentration for each group in both treatments, comparing the trehalose levels of the diabetic flies to the wild type flies as shown in **Graph 2**. We conducted the T-test using a 95% confidence interval. **Graph 2** shows the trehalose levels of the diabetic and wild type *Drosophila* in both the green tea and media only with the 95% confidence intervals. Because the confidence intervals of the diabetic flies and wild type flies in media overlap, we do not have enough data to detect a difference in trehalose levels between the diabetic and wild type *Drosophila*. The trehalose concentration of the diabetic flies in green tea extract was 0.0506 mg/mL higher than the wild type flies in green tea extract. Because the confidence intervals of the wild type flies in media and green tea, however, do not overlap, we conclude that the trehalose levels of the wild type flies in green tea are statistically lower than the wild type flies in media only. This shows that the green tea extract resulted in overall lower trehalose levels in wild type flies only.

DISCUSSION

With approximately 500 million cases worldwide, the global prevalence of type 1 diabetes continues to rise.[6] Current treatment aims at maintaining normal blood sugar levels through regular monitoring, insulin therapy, diet, and exercise. Before insulin was available, the only treatment for diabetes was to restrict carbohydrates in the diet and keep free glucose in the blood to a minimum.[10]

This study focused on testing the effects of green tea extract on type 1 diabetic flies. We compared the trehalose levels of diabetic and wild type flies in standard *Drosophila* media and in media infused with green tea extracts. We hypothesized that the trehalose levels in diabetic flies with a green tea supplemented diet would be significantly lower than the diabetic flies in normal media. We found no statistical difference between the average trehalose concentration of the diabetic flies in media only as compared to those in the green tea extract, with a difference of only 0.002 mg/mL. When using a 95% confidence interval, the values overlapped, signifying no statistically significant difference between the two samples. This could be related to the small sample size of eight diabetic fly samples in the green tea extract. We were unable to collect more diabetic flies in the green tea because there was significantly reduced reproduction in the vials with no eggs or larvae visible even after several weeks.

Although there was no clear difference in the trehalose concentrations of the diabetic flies in media only compared to those in the green tea extract, the trehalose concentration of the wild type flies in the green tea was significantly lower than the wild type flies in media only, with an average decrease of 0.0506 mg/mL. Using a 95% confidence interval, we found these results to be significantly different.

In similar research, Richard Anderson, a biochemist at the USDA's Beltsville Human Nutritional Research Center, reported that EGCG influences the primary way that glucose is released by the liver. He stated that EGCG may help diabetics by mimicking the action of insulin in inhibiting the liver's release of glucose, thus lowering the overall circulating blood sugar level.[14] It is possible that the EGCG affected the *Drosophila* in similar ways. The wild type flies in the green tea exhibited lower circulating free trehalose levels which may have been caused by EGCG mimicking signals produced by insulin and inhibiting the release of trehalose into the blood. In the diabetic flies, their trehalose levels were naturally higher than in the wild type flies, but EGCG did not affect the circulating free trehalose the same way. These results cannot be explained based on research from the previous study and suggests that EGCG affects the trehalose levels through some other mechanism.

Another study found that one cup of green tea inhibited the activity of amylase by 87%, which is the enzyme required to break down starch into simple sugars that can be absorbed in the blood stream.[5] This study also supports our findings in the wild type flies, but does not help us understand the lack of change observed in diabetic flies. In the end, we found that while the green tea seemed to have no effect on the trehalose levels of the diabetic flies, it did have a statistically significant effect on lowering the trehalose levels in wild type flies.

Throughout our experiment, there were many errors that could have affected our results. With a larger sample size, however, we might have observed a difference in the diabetic flies. The lowered reproduction rate of the diabetic flies in green tea resulted in a lower sample size and a longer time before the flies were collected after hatching from larval form. This gap between the time the *Drosophila* were larvae, eating the media, to adult, could have nullified the effects of the EGCG on the trehalose levels. This limited the number of samples we were allowed to test, and the small sample sizes yielded results statistically insignificant. Another possible source of error was the amount of trials we were able to conduct. Due to limited time, we were only able to conduct two trials of each group in the green tea and media. If we had time to conduct more trials and collect more flies from each group in the green tea and media only, we would be able to confirm the data that we have collected.

The next step for our research would be to rerun the experiment, conducting more trials in order to collect more samples, and collect flies as they hatch or test the trehalose levels of the larvae. Another step for our research would be to test the effects of green tea on type 2 model of *Drosophila*, concerning the effects of EGCG on decreasing the insulin resistance.

ACKNOWLEDGMENTS

We would like to thank Cayla Jewett and Kelsey Hazegh, graduate students from the Molecular Biology Program at the University of Colorado, Denver for mentoring us through the scientific process, for their technical assistance, for donating the Dilp2>Gal4/CyO flyline we used to create our diabetic model, and for contributing meaningfully to the design and methodology of our research. We would also like to thank Matt Bernstein of Thermo Fisher Scientific for sharing his expertise and instructing us on the functions of the nanodrop. We extend our gratitude to Brian Olcott and Sandra Tavalario for funding our research, as well as those who anonymously donated. We thank Tom Dillon for aiding in our research design and acknowledge the contributions of Amy England and Jim McClurg for producing the video featured on our webpage. A special thank you to Brian Winkelman for managing and designing our website as well as his involvement in this publication. Last but certainly not least we appreciate the editorial support of Wendy Lerolland. The present work was also supported in part by Rock Canyon High School and the Douglas County School District.

REFERENCES

1. Baker, K. D., & Thummel, C. S. (2007). Diabetic larvae and obese flies-emerging studies of metabolism in *Drosophila*. *Cell Metabolism, 6*(4), 257-266. DOI:10.1016/j.cmet.2007.09.002
2. Bogdanski, P., Suliburska, J., Szulinska, M., Stepien, M., Pupek-Musialik, D., & Jablecka, A. (2012). Green tea extract reduces blood pressure, inflammatory biomarkers, and oxidative stress and improves parameters associated with insulin resistance in obese, hypertensive patients. *Nutrition Research, 32*(6), 421-427. DOI:10.1016/j.nutres.2012.05.007
3. Carlson, S., Castro, M., & Moreland, P. (2014). Type 1 Diabetes. Mayo Clinic. Retrieved 2016, October 5. [Web]
4. Dus, M., Min, S., Keene, A. C., Lee, G. Y., & Suh, G. S. (2011). Taste-independent detection of the caloric content of sugar in *Drosophila*. *Proceedings of the National Academy of Sciences,108*(28), 11644-11649. DOI:10.1073/pnas.1017096108
5. Hara, Y., Honda, M. (1990). The inhibition of α-amylase by tea polyphenols, agricultural and biological chemistry, 54:8, 1939-1945. DOI: 10.1080/00021369.1990.10870239
6. Kostic, A., Gevers, D., Siljander, H., Vatanen, T., Hyötyläinen, T., Hämäläinen, A., . . . Xavier, R. (2015). The dynamics of the human infant gut microbiome in development and in progression toward type 1 diabetes. *Cell Host & Microbe,17*(2), 260-273. DOI:10.1016/j.chom. 2015.01.001
7. Knowler, W.C., Barrett-Conner, E., Fowler, S.E., Tuomilehtoj, J., Lindstom & J., Eriksson, J. (2014). More than 29 million Americans have diabetes; 1 in 4 doesn't know. National Center for Chronic Disease Prevention and Health Promotion. Retrieved 2016, September 23. [Web]
8. Kroon, E., Martinson, L. A., Kadoya, K., Bang, A. G., Kelly, O. G., Eliazer, S., . . . Baetge, E. E. (2008). Pancreatic endoderm derived from human embryonic stem cells generates glucose-responsive insulin-secreting cells in vivo. *Nature Biotechnology, 26*(4), 443-452. DOI:10.1038/nbt1393
9. Manning, G. (2008) A quick and simple introduction to *Drosophila melanogaster*. Ceolas. Retrieved 2016, October 5. [Web]
10. Mazur, A. (2011). Why were "starvation diets" promoted for diabetes in the pre-insulin period. BioMed Central. DOI: 10.1186/1475-2891-10-23
11. Stanford researchers create diabetic fruit flies in lab. (2002). American Association for the Advancement of Science. Retrieved 2016, November 18. [Web]
12. Statistics about diabetes. (2016). Centers for Disease Control and Prevention, National Institutes of Health, American Diabetes Association. Retrieved 2016, October 7. [Web]
13. Stem Cells and Diabetes. (2001). National Institutes of Health. Retrieved November 19, 2016. [Web]
14. Tea gives big boost to insulin. (2002). UPI Science News. Retrieved March 19, 2017. [Web]
15. Tennessen, J. M., Barry, W. E., Cox, J., & Thummel, C. S. (2014). Methods for studying metabolism in *Drosophila*. *Methods, 68*(1), 105-115. DOI:10.1016/j.ymeth.2014.02.034
16. Waltner-Law, M. E., Wang, X. L., Law, B. K., Hall, R. K., Nawano, M., & Granner, D. K. (2002). Epigallocatechin gallate, a constituent of green tea, represses hepatic glucose production. *Journal of Biological Chemistry, 277*(38), 34933-34940. DOI:10.1074/jbc.m204672200

ABOUT THE AUTHORS

Pictured: In the middle are Kelsey Hazegh (left) and Cayla Jewett (right), our mentors from the molecular biology program at the University of Colorado, Denver. Also pictured is Jason Olcott (far left) and Sean Nasseth (far right).

This year challenged us in many ways. The opportunity to conduct graduate level research empowered us to grow as people and further understand and appreciate science as it relates to our lives. Gaining exposure to the entire scientific process including pitching a proposal to a panel, acquiring funding, as well as designing and conducting research focused around type 1 diabetes enriched our prospects of pursuing research in college. Jason plans to study biology and Sean plans to study industrial design. Our journey through this project was initially inspired by Jason's own battle with type 1 diabetes. He was diagnosed when he was fourteen, which reflects in our strong emotional investment for our work. Designing a feasible project, organizing and managing the workload, accounting for unexpected turns associated with the green tea and other variables, and acquiring the necessary equipment to get the data were a handful of the many obstacles we had to climb over. All of this and more revealed to us the sheer reality that research is a full-time job that requires diligent time and effort with a keen attention to detail. We were fortunate to have considerable support from our teachers, mentors, donors, parents, and of course, each other. We attribute the aptitude of our project to their investment. Utilizing the resources around us proved invaluable when faced with the challenges we had and our research experience taught us the importance of perseverance in ultimately turning around failure to achieve a goal. The promise of this course sets precedent that for even a couple of high school students, there is no limit.

A comparison between three different protocols for anesthetizing *Drosophila melanogaster* in a high school classroom

S. Harikrishnan and S. L. Fordham

Department of Science, Principles of Experimental Design in Biotechnology, Rock Canyon High School, Highlands Ranch, Colorado, USA

In high school laboratories, *Drosophila melanogaster* (fruit fly) is a common model organism used to help students understand genetic crosses or mutations. The predominant method for anesthetizing *Drosophila* in high school labs involves the use of a chemical anesthetic known as FlyNap®, which contains triethylamine, a toxic and potentially hazardous chemical. *Drosophila* are required to be anesthetized for students to score, sex, and separate virgin females from males in the process of performing genetic crosses. This project compared of ease of use, safety, and the cost of three different anesthetic methods: FlyNap®, CO_2 gassing, and freezing techniques with *Drosophila*. While the use of FlyNap® does have benefits, such as a short initial anesthetization time and keeping the flies asleep for a long period of time, it scored the lowest from students and teachers due to ease in killing flies and the harmful health effects of working with FlyNap®. Students preferred CO_2 gassing over FlyNap®. Freezing was found to be the most efficient method to use in a high school classroom setting, as it required the least amount of time to set up, allowed the flies to remain asleep for a long period of time, was the most cost efficient method, and required the minimum safety precautions. Teachers preferred this method of the three anesthetization methods. The recommended method of anesthetization in a high school classroom is freezing for its benefits over FlyNap® and CO_2 gassing.

Drosophila melanogaster is one of the most commonly used model organisms in both high schools and research institutions to study genetics and development to prepare students interested in pursuing careers in biological sciences.[7] Students who take classes involving work with organisms such as *Drosophila* have generally never worked with model organisms and are not proficient with handling, anesthetizing and observing these organisms. Without practice, students tend to make multiple errors and sterilize or kill the flies due to over-anesthetization. Training and efficiency is required for students to follow and execute the protocols efficiently.[4]

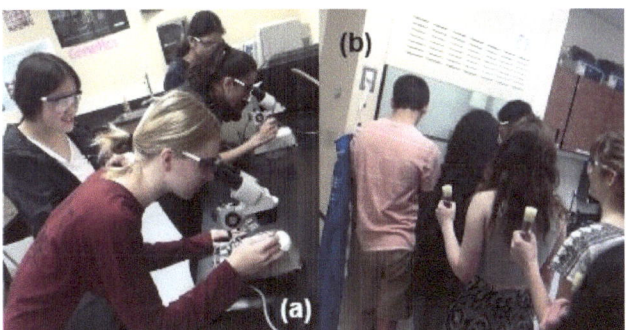

Picture 1: (a) This is a picture of a student sorting and sexing flies under the stereomicroscope. (**b**) A visual representation of students waiting in line for the fume hood which is used to gas flies using FlyNap®. An entire class of students waiting in line to gas flies can take up to 45 minutes.

Drosophila is the ideal organism for conducting experiments that test genetics, evolution, biochemistry and behavior. They are easy to raise in labs because they are cheap, small in size, occupy a small amount of space, and have simple food requirements. Females lay over 400 eggs, and have a life cycle of 10-14 days.[1] Experiments crossing *Drosophila* with different phenotypes help students visualize each cross and better understand genetic inheritance. *Drosophila* have been used for over 100 years in genetic research and their traits are easily observable using a stereomicroscope. Some commonly researched genetic traits in high schools are curly wings (Cy), apterous (ap), bar eyes (B), and white eyes (w).[6] Because they are invertebrates, legal constraints do not prevent their use in the lab. *Drosophila* are easy for new learners due to culture and handle. An abundance of literature provides strategies and information to instructors to achieve the ideal results with their students.[1]

A variety of protocols is used to anesthetize *Drosophila* in biological laboratories. Among them, gassing with FlyNap®, gassing with CO_2, and freezing are the most commonly used protocols. Although they are all used to achieve the same results, these methods are distinctly different. FlyNap® is the anesthetic agent a majority of high school instructors use. FlyNap® uses chemicals such as triethylamine to anesthetize *Drosophila*. A benefit to this method is it takes less than five minutes to knock out the *Drosophila* and keeps adult flies asleep for approximately 50 minutes at a time. This is sufficient time for inexperienced students to observe the flies under a stereomicroscope, record observations, and score results (**Pic. 1**).[5] However, FlyNap® contains triethylamine, which can cause headaches and dizziness at high concentrations. Triethylamine can be toxic to humans through inhalation or contact with skin, as it targets the lungs and liver if it is ingested into the body in high concentrations.[5] Studies have shown that triethylamine cannot only harm students, but can also cause cardiac physiology in *Drosophila*.[3] Instructors have received complaints from students about dizziness and headaches after working with FlyNap®.[5] In addition, it can be especially dangerous for instructors who are exposed to FlyNap® fumes throughout the entire day. My research will evaluate the benefits of a long sleep time compared to potential risks associated with the use of FlyNap® in a classroom.

Alternative methods for FlyNap® include gassing with CO_2 and freezing under -18°C for several minutes.[5] Gassing with CO_2 does not involve harmful chemicals such as triethylamine and does not put the students at risk for health issues. The concentrations of CO_2 released during the anesthetization of *Drosophila* are minimal and safe. Many labs use CO_2 to anesthetize the flies and then lay the flies on a CO_2 pad. Traditional methods of CO_2 anesthetization require exposure to CO_2 for a prolonged period of time because the *Drosophila* rest on a pad which releases a steady supply of CO_2 at fairly high concentrations (**Pic. 2**).[1] If it was not used in ventilated zones, high concentrations of CO_2 could cause severe dizziness and headaches.[1]

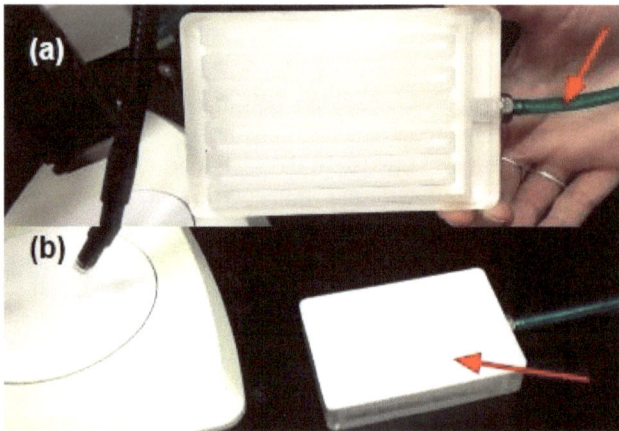

Picture 2: (a) This is a picture of a CO_2 pad which is most commonly used in college laboratories. The CO_2 gas is delivered to the flies through the green tube to the right of the pad. **(b)** For anesthetization, the vial containing the flies will be upturned onto the pad until the flies fall asleep.

Although the use of CO_2 may be less harmful than triethylamine, prolonged periods of exposure to this colorless and odorless gas at high concentrations has resulted in health hazards for the organism and researchers studying the organism.[1] Behavior, reproduction and neurology of young *Drosophila* has been affected by prolonged exposure to high concentrations of CO_2.[7] Studies have shown a decrease in reproductive activities and an increase in copulation latency of *Drosophila* even after they have been in recovery for over 20 hours.[1,2] In my research, I used the Carolina Biological Carbon Dioxide Anesthetization kit containing low concentrations of CO_2 to put the flies to sleep and then transferred then into a petri dish lined with filter paper atop a cold pack. This method used a small concentration and a small period of exposure to CO_2. Instead of the traditional CO_2 tank, I used an Alka Seltzer tablet in water as the source of CO_2 (**Pic. 3a**).

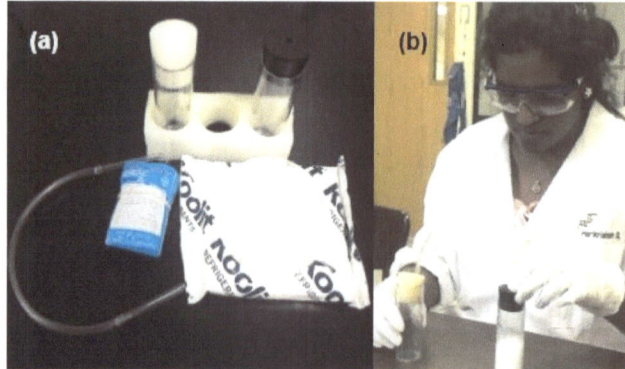

Picture 3: (a) This picture shows the Carbon Dioxide Anesthetizing Kit from Carolina Biological. **(b)** Harikrishnan gassing *Drosophila* using the kit.

The third method of anesthetization that I evaluated was freezing at -20° C. Freezing is used in a small number of high schools but dominates in research facilities that conduct genetic experiments with *Drosophila*. Among the three methods, freezing posed the least potential harm to students when performed correctly because no harmful substances or gases were used. Freezing not only kept the students away from chemicals, but also protected the *Drosophila* from exposure to chemicals that could sterilize them.[1] The flies were placed into a freezer at -20° C in an empty culture vial until they fell asleep. They were then moved out of the freezer onto a petri dish which had been placed atop a cold pack.[7] Unlike with CO_2 and FlyNap®, cooling does not affect the behavioral traits of *Drosophila*.[5] However, extensively anesthetized and unconscious *Drosophila* are not achieved without some harm done to the organism.

While freezing does not pose any dangers chemically, it reduces fly activity physically. Studies have shown disruptions in fruit fly memory because of the use of freezing.[2] While memory loss is not a major problem for the type of research conducted in most high school labs, when the flies are set on metal plates or petri dishes for observation, condensation is a potential fatal problem. The wings of *Drosophila* can get soaked if condensation occurs. Male *Drosophila* with wet wings are unable to sing courtship songs which lowers mating behavior.[2,7] This issue can be alleviated with the use of filter paper placed on top of the metal plates or petri dishes to absorb the moisture and protect the fragile wings of the *Drosophila*. In this experiment, I evaluated the best anesthetization method among FlyNap®, CO_2 gassing and freezing for use in a high school classroom by comparing the cost, safety, and ease of use of each protocol.

METHODS

Wild type *Drosophila melanogaster* were obtained from the Rock Canyon High School Biotechnology Lab. Formula 4-24® Blue Instant *Drosophila* Medium (17-3210) from Carolina Biological was used to culture the flies. All other materials required for the project, such as FlyNap® and culture vials, were provided by Rock Canyon High School. The flies were anesthetized using three different protocols:

FlyNap® gassing, CO_2 gassing and freezing. Cost, safety, each method were performed and measured for time to sleep, time to wake, and time to death. Each assay was performed separately with different flies. In all three of the anesthetic treatments, 25-50 flies were used so statistical analysis tests could be performed.

Anesthetization Protocols

For all three protocols, prior to being anesthetized, the flies were transferred into an empty culture vial using a funnel (**Pic. 4**). For both the CO_2 and freezing trials, when all the flies were rendered unconscious, they were transferred to an empty petri dish lined with filter paper, as they would wake up in 20-30 seconds if they were not cooled right after anesthetization (**Pic. 5**). FlyNap® and CO_2 gassing protocols provided by the Carolina Biological Supply Company were used as a reference when anesthetizing the flies.[4]

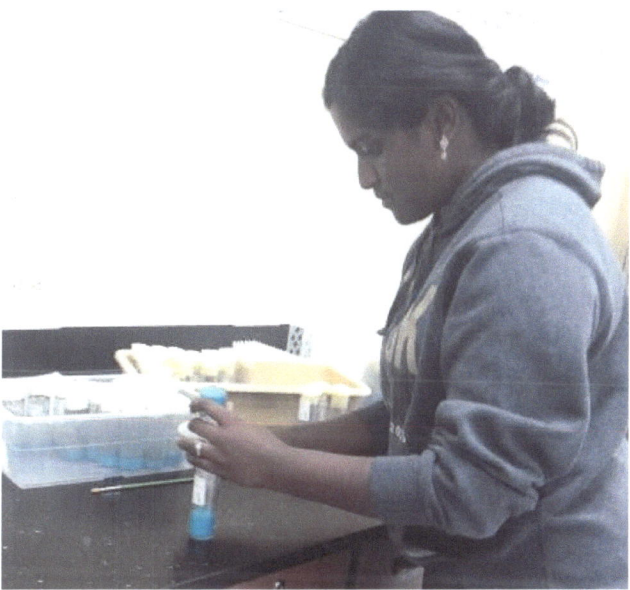

Picture 4: Harikrishnan divides flies equally and transfers them into multiple culture vials.

FlyNap®

The FlyNap® gassing protocol was performed under the fume hood. After the flies were transferred from their culture vials into the empty culture vial, the anesthetic wand was then dipped into the FlyNap® solution and the absorbent end of the wand was inserted directly below the foam plug into the vial containing the flies. The vial was then tapped several times on a flat surface to ensure that the flies did not stick onto the anesthetic wand. Once all of the flies stopped moving, they were removed from the gassing chamber and transferred onto an index card to be observed and counted using a stereomicroscope.

CO_2 Gassing

After the flies were transferred into an empty culture vial, 15 ml of water was added to a second empty culture vial with one Alka Seltzer tablet. This culture vial was plugged with a rubber stopper connected to a tube leading, which

was immediately inserted into the vial containing the flies (**Pic. 3b**). Once all the flies stopped moving, they were transferred onto a petri plate lined with filter paper on top of an ice pack to be observed and counted using a stereomicroscope.

Freezing

After the flies were transferred into an empty culture vial, they were placed inside a freezer at -20°C until they were no longer moving. The flies were then immediately transferred onto a petri dish lined with filter paper and were observed and counted under a stereomicroscope.

Time Tests

To test time to sleep, the flies were anesthetized until they were unconscious. This was performed to obtain an average sleep time for the flies with each method. With the averages, instructors will be able to give students an appropriate amount of time which the students can use to anesthetize the flies.

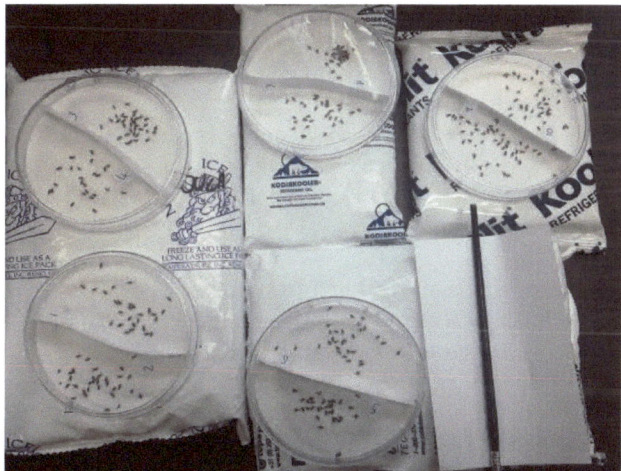

Picture 5: A photo of flies in their designated petri dishes for obtaining time to wake. The vials are numbered according to the increments in time to sleep.

The second time test was time to wake, which was the time it took for the flies to wake up after being fully anesthetized. This was performed to determine the amount of time students had to work with the flies before they woke. After complete anesthetization, time was recorded when the first fly regained normal. In addition, the time it took for half of the flies to become active was recorded along with the time it took for all of the flies to become active. If the flies did not wake after two hours, the time was not recorded because two hours is a sufficient time for the students to complete all observations and record data.

For the final time test, time was recorded when flies died due to over-anesthetization. The flies were first anesthetized using the average sleep time, as measured in the time to sleep trials and then they were exposed to additional anesthetization in 15-30 second increments. Separate flies were used for each time interval increment and they were left overnight to recover in a culture vial with media. The next day I observed the vials and recorded the number of flies awake and dead. Each treatment was assessed using a different group of wild type flies to minimize errors due to repeated anesthetization.

When rating each method on the time tests, cost, and safety, a scale of 1-5 was used. For time to sleep, a score of 5 represented a method where flies would fall asleep within a 5 minute anesthetization period and a score of 1 represented anesthetization that prolonged for over 10 minutes. All the methods took less than five minutes to anesthetize the flies which makes each a viable option for classroom use. For time to wake, a score of 5 meant the first fly would wake up no sooner than 50 minutes after anesthetization and a score of 1 represented that the first fly woke before 10 minutes. The wake time was not recorded past two hours for any method because a class period usually will not last longer than two hours and it is a sufficient time for students to make observations. For time to death, a score of 5 represented no fly death with 1 minute and 30 seconds or more of over-anesthetization and a score of 1 represented fly mortality at 30 seconds or less of over-anesthetization. For cost, the three methods were compared to each other and rated on a scale of 1-5. The least expensive method received a 5 and the most expensive method received a 1. For safety, a score of 5 represented a method with the least amount of safety precautions and had the least amount of side effects to the student and teacher, while a score of 1 indicated a method that posed significant safety concerns.

Ease of Use

To measure ease of use, I had 26 volunteer students and two Biotechnology teachers perform and rate each of the three protocols. Because all methods required the transfer of flies into a separate culture vial, I performed that step and only had them perform the actual anesthetization. Ease of use was evaluated by students and teachers through a survey where they rated each method on its compatibility with a high school classroom and selected an anesthetization method overall.

RESULTS

In this experiment, I evaluated the best anesthetization method for *Drosophila* comparing FlyNap®, CO_2 gassing, and freezing for use in a high school classroom by comparing the cost, safety and ease of use for each protocol. I conducted three trials for time to sleep, time to

wake, and time to death for each anesthetization method.

FlyNap® received a score of 5 for time to sleep because its average anesthetization time was 1 minute and 45 seconds (**Graph 1**). However, in a classroom setting, it received a score of 1 for time to sleep because it took approximately 30-40 minutes for a class of 30 students to get a turn at the fume hood (**Pic. 1**). It received a score 5 for time to wake because the average time to wake for FlyNap® after anesthetization was approximately 60 minutes (**Graph 2**). It received a score of 1 for time to death because FlyNap® had the highest risk of fly death, which can occur after only 30 seconds of over-anesthetization (**Graph 3**). FlyNap® received a score of 3 for cost. The initial cost includes the purchase of the FlyNap® anesthetization kit which costs approximately $14, however the amount of FlyNap® in this kit may not be sufficient for one class throughout the experiment. Additional FlyNap® would most likely have to be purchased to supplement this kit. The recurring cost would only include the FlyNap® solution. A bottle of 100ml FlyNap® anesthetic costs approximately $69 from Carolina Biological and would be sufficient for several classes over several years. It scored a 3 for safety because it is best performed under a fume hood, requires safety precautions such as eye protection, and can give students and teachers headaches.

CO_2 gassing received a score of 5 for time to sleep because the anesthetization time was the shortest for CO_2 gassing with an average of 40 seconds (**Graph 1**). In a classroom setting students could anesthetize the flies at their stations with a short anesthetization time. It received a score of 5 for time to wake because the flies that were anesthetized using CO_2 gassing did not wake up by the end of the two hour period and stayed asleep for as long as they were on the ice pack (**Graph 2**). It received a score of 4 for time to death because a few flies were over-anesthetized with one Alka Seltzer tablet. CO_2 gassing received a score of 1 for cost because it is the most expensive method of all the three methods. The carbon dioxide anesthetization kits can either be purchased from Carolina Biological at a cost of approximately $14 for each kit or made from materials from a hardware store or found in the lab. The amount of kits that would have to be purchased for a class varies depending on the structure of the groups and amount of students present in the class. For a class of 30 students who are working in pairs, 15 kits would have to be purchased at

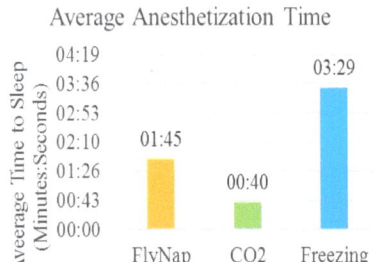

Graph 1: A visual representation of the average anesthetization time for the three methods.

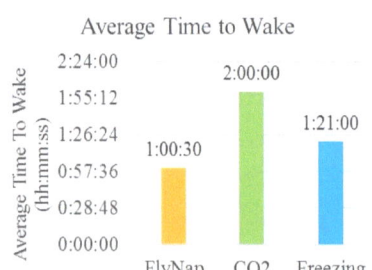

Graph 2: This graph shows the average time for flies to wake after anesthetization.

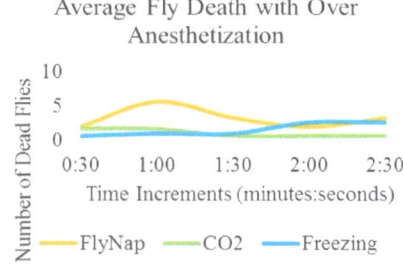

Graph 3: The flies were fully anesthetized and after anesthetization the average amount of fly death with the over anesthetization of 30 second increments for each method has been graphed.

a cost of approximately $200. Filter paper also has to be purchased for lining the petri dishes which costs approximately $15 per box of 100 papers and a pair of students will use about two petri dishes. Any type of ice packs can be purchased, if they are not already available, for keeping the flies cool after anesthetization. The Polar Tech 20oz Foam Freeze pack can be purchased for $14 and it contains 12 ice packs. The carbon dioxide anesthetization kit, ice packs, and filter paper are initial purchases which total to $300. In addition, Alka Seltzer tablets would have to be purchased at approximately $6 for a box of 36 tablets. This would be enough for 2-3 teams of 2, so the reoccurring cost would be paper are initial purchases which total to $300. In addition, Alka Seltzer tablets would have to be purchased at approximately $6 for a box of 36 tablets. This would be enough for 2-3 teams of 2, so the reoccurring cost would be $90 per class. It received a score of 5 for safety because no harmful chemicals are used and there are no safety risks to the students.

Freezing received a score of 5 for time to sleep even though it took the longest time to anesthetize the flies, averaging 3 minutes and 29 seconds (**Graph 1**). In a classroom setting, however, freezing would be the most efficient method because the vials of an entire class can be gassed in under five minutes. It received a score of 5 for time to wake because the average time to wake for the flies that were anesthetized using freezing was 1 hour and 21 minutes (**Graph 2**). Freezing received a score of 5 for time to death because it had the lowest fly death at 30 seconds, which would be the most common amount of over-anesthetization time seen with students. Freezing allows for an over-anesthetization period of 1 minute and 30 seconds before the first fly death. This method would therefore be best for inexperienced students. Freezing received a score of 5 for cost and safety because you can use a regular freezer. Ice packs and filter paper have to purchased but they can be reused after the first purchase and there are no safety hazards. For use with a classroom of 30 students, the initial cost of freezing would be a total cost of approximately $50.

Which Method Do You Prefer Overall?

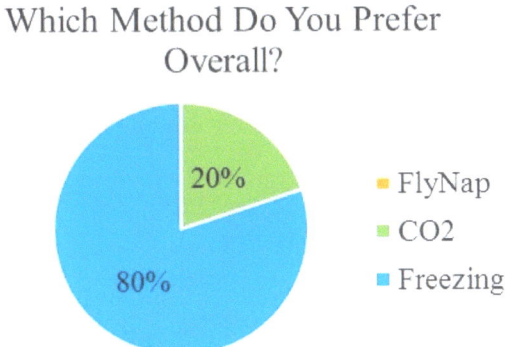

Graph 4: A visual representation of the anesthetization method students preferred overall in a classroom setting.

For calculating ease of use, 26 students and two teachers were surveyed on which method they preferred after performing each of the three anesthetization protocols. All of the students and teachers preferred freezing or CO_2 gassing over FlyNap® (**Graph 4**). 20% of the students preferred to use CO_2 gassing over freezing, while 80% of students and all of the teachers preferred freezing (**Graph 4**). The 20% of students who preferred CO_2 gassing, thought it was entertaining to watch the Alka Seltzer bubble inside the vial and to witness how fast the flies fell asleep. Others said that they preferred the anesthetization time of under one minute and that they could remain at their table rather than going to the fume hood, surrounded by the smell of FlyNap®. Most students and all teachers found freezing as the easiest method to perform and teach because students only had to transfer their flies from the culture vial into an empty culture vial. The anesthetization process could be completed in under 10 minutes for an entire class of 30 students, while FlyNap® can take up to 40 minutes due to the long lines to get into the fume hood.

DISCUSSION

Drosophila is one of the most common model organisms used to study genetics and mutations in high school laboratories. The method most often used to anesthetize the flies in a high school setting is FlyNap® contains triethylamine, a potentially hazardous substance, that in high concentrations can cause bad headaches and even migraines in students and teachers. In this research, I compared CO_2 gassing and freezing anesthetization methods to FlyNap® by averaging time to sleep, time to wake and time to death due to over-anesthetization for each of the methods. I evaluated the ease of use, cost and safety for these methods and analyzed this data to determine the best method for use in a high school laboratory.

As shown in the results, FlyNap® scored high for time to sleep and time to wake but had the highest mortality due to over-anesthetization of all the methods. Even though FlyNap® had a short amount of anesthetization time, it takes much longer in a classroom setting due to the restricted access of students in the fume hood (**Pic. 1b**). Not only does FlyNap® take longer to perform in a classroom setting, it requires equipment such as a fume hood, which not all high schools may have access to. FlyNap® is also not very cost-efficient as the cost of a FlyNap® anesthetization kit is $14 and the solution has to be purchased every year, as it is used throughout the experimentation period by multiple classes. One of the biggest drawbacks of FlyNap® was its lack of flexibility with over-anesthetization. Fly death started at 30 seconds of over-anesthetization and it had the highest mortality rate of all three methods (**Graph 3**). FlyNap® was the most difficult method because of the short time frame to anesthetize the flies before killing them. It received the lowest rating from students and teachers as they found it easy to kill their flies and the smell of FlyNap® was unpleasant and many experienced headaches with its use (**Graph 4**).

CO_2 gassing similarly scored high on time to sleep and time to wake due to its very short anesthetization time and long time the flies remained asleep after anesthetization.

Killing the flies with over-anesthetization from the Alka Seltzer tablet did not occur easily because the tablet stopped reacting inside the gassing chamber at around five minutes and produced a finite amount of CO_2 gas. The difficulty with this method was that the flies had to be taken out of the gassing chamber and placed on the ice pack very close to the average anesthetization time of 40 seconds, otherwise they would begin to wake up before the students could collect data. The biggest drawback with CO_2 gassing is the cost. The startup total costs could be as high as $350 and the reoccurring yearly cost for one class as high as $60.

Freezing received the highest rating overall. With this method, flies can be anesthetized by placing the vial of flies inside a -20°C freezer for 3-4 minutes. Some flies still survived with 20 minutes of over-anesthetization. It was shown to be the best method to use with a classroom of inexperienced students because the flies could be over-anesthetized for 1 minute and 30 seconds until they started to die (**Graph 1**). Freezing also had the lowest overall fly death by over-anesthetization among the three methods. It is the most efficient method to use in the classroom because students are only required to transfer their flies into an empty culture vial. The instructor can then take all of the vials and anesthetize them in less than five minutes. Anesthetization using freezing is also the cheapest method because there are no recurring purchases. For use with a classroom of students, the initial cost of freezing would include ice packs, petri dishes and filter paper at a total cost of approximately $50. Freezing does not pose a safety risk because the students and teachers do not come in contact with any harmful chemicals.

To assist schools who need more inexpensive options, three other anesthetization methods were tested: vinegar and baking soda, ice cubes and crushed ice. Vinegar and baking soda is a plausible option for anesthetization but does not work if done in a culture vial as it overflows and drowns the flies in the gassing chamber. Ice cubes were not an efficient anesthetization method for those without a freezer because the space between each ice cube was too great and it did not anesthetize any flies. Crushed ice was an effective alternative to freezing and it works very efficiently. A vial of flies can be fully placed in the crushed ice until the flies are asleep and can be placed on an ice pack to be observed. For a classroom setting, a bucket of crushed ice can be used as an alternative to the freezer.

Freezing was the best anesthetization method overall for use in a high school classroom. It is the most time efficient, flexible, and has room for student error, is the easiest to perform, is the least expensive, and does not have any potentially harmful effects to students and teachers. It has a short anesthetization time and results in flies that stay asleep for a long period of time. I recommended that high school teachers using *Drosophila* use freezing as their anesthetization method.

Limitations of this experiment were that freezing and CO_2 methods were not tested with a classroom of 30 students who were collecting newly hatched virgin flies. The data may change with younger flies rather than the heartier adults I used for my data collection. The data collection was also performed by an experienced student rather than the inexperienced students who will be performing the experiments.

Further research can be conducted to assess the efficiency of the anesthetization methods on flies at different stages in development. Most genetic experiments in high schools use virgin flies which may be more susceptible to having problems with one of these anesthetization methods. Further research can explore if newly hatched virgins are more prone to die from CO_2 gassing and/or freezing. Another option for further research can be testing CO_2 gassing and freezing on *Drosophila* with common mutant traits used in genetic experiments to see whether the different anesthetization methods have an effect on the phenotypes of the flies.

ACKNOWLEDGMENTS
I would like to thank Uma Venkitanarayanan for peer reviewing and supporting me with the experimental design of this research. I also thank, Stephanie Bonney and Santiago Fregoso, graduate students at the University of Colorado, Denver. Thanks to Jim McClurg and Amy England for filming and editing my proposal video. I would also like to thank Tom Dillon for his support with my experimental design. I would also like to especially thank Harikrishnan Kandaswamy for donating to my project. Thank you, Susanne Petri and Nikki Dobos, for allowing me to share your classroom and lab space throughout this experiment and Wendy Lerolland for providing continuous feedback and editorial assistance. Thank you to Bryan Winkelman, who played a tremendous supporting role in this course, from formatting this article to reading my blog posts and maintaining our website. Last, I would like to thank Rock Canyon High school for providing the supplies, equipment, and laboratory space that made this project possible.

REFERENCES
1. Artiss, T., & Hughes, B. (2007). Taking the headaches out of anesthetizing *Drosophila*: a cheap and easy method of constructing carbon dioxide staging. *The American Biology Teacher*, 69(8), e77-e80.
2. Barron, A.B. (2000). Anesthetizing *Drosophila* for behavioural studies. *Journal of Insect Physiology* 46, 439-442.
3. Chen, W., & Hillyer, J. F. (2013). FlyNap (triethylamine) increases the heart rate of mosquitoes and eliminates the cardioacceleratory effect of the neuropeptide CCAP. *PloS one*, 8(7), e70414.
4. Flagg, R. O. (2005). *Carolina Drosophila Manual*. Burlington: Carolina Biological Supply Company. [Print]
5. Ratterman, D. M. (2003). Eliminating ether by using ice for *Drosophila* labs. *Tested Studies For Laboratory Teaching*, 24, 259-265.
6. Roberts, D. B. (2006). *Drosophila melanogaster*: the model organism. *Entomologia experimentalis et applicata*, 121(2), 93-103. [Web]
7. Wen-hui Q., Tong-bo Z., & Da-Xiang Y. (2015). A modified cooling method and its application in *Drosophila* experiments. *Journal of Biological Education* 49, 302-308. [Web]

ABOUT THE AUTHORS

Pictured: From left to right, Shreemathi Harikrishnan and my mentor Shawndra Fordham.

I have been interested in biology for the majority of my life. I had planned on pursuing veterinary sciences but I have changed my mind to pursue Biotechnology because of the thorough enjoyment I experienced during this course. Over the course of the year, I have learned many valuable lessons that will help me create meaningful opportunities in the future. Managing time wisely and having patience for a process to complete became a frustration during some critical points of the experimental period. But I have gained strength from those experiences when I learned to push myself, as well as learn my own limits. This memorable course has taught me resilience unlike any other course I have experienced in my high school career. I was able to advance to heights that I had never imagined being in as a senior in high school.

Expression of PPARG in precancerous hyperplastic skin cells in relation to the presence of Aurora Kinase A protein and the p53 gene

M. Kumar, W. M. Shannon, S. A. Sharma, and S. L. Fordham
Department of Science, Principles of Experimental Design in Biotechnology, Rock Canyon High School, Highlands Ranch, Colorado, USA

Squamous cell carcinoma is a form of skin cancer in which the epidermal layer of skin develops abnormal cells as a result of environmental mutagens. Changes in Aurora Kinase A protein and p53 gene, which generally regulate healthy cell proliferation and tumor suppression, have been linked to the expression of peroxisome proliferator-activated receptor gamma (PPARG), and therefore may indicate the onset of squamous cell carcinoma.[6] Thus, using immunohistochemistry, we tested the expression of PPARG in hyperplastic, or precancerous, skin cells in order to determine whether its presence could serve as an indicator of the onset of skin cancer. Using this procedure, we performed multiple trials; however, after viewing the stained slides under a microscope, we saw there was abundant amount of background noise present that interfered with the viewing of the actual staining. Furthermore, we noticed that the staining we were able to decipher was faint, even in slides that we expected to show a strong signal. Our results reflected that the antibodies used in this experiment were not optimal in detecting the protein PPARG, and we therefore were unable to draw conclusions as to whether the presence of PPARG could serve as an indicator for squamous cell carcinoma.

Squamous cell carcinoma skin cancer is the second-most contracted type of skin cancer. Primarily, it is a result of exposure to ultraviolet rays which damage the cells in the epidermal layer **(Fig. 1)**.[5] The epidermis creates a protective barrier against bacteria and viruses and helps regenerate keratinocytes; however, when it is exposed to excessive sun and UV radiation, abnormal cells begin to form rapidly, causing actinic keratosis to form and further develop into squamous cell carcinoma.[4] Early detection of squamous cell skin cancer is important for prevention of metastasis; therefore, understanding signs of early detection could prevent the spread of cancer. Since the Aurora Kinase A protein and the p53 gene are linked to the formation of actinic keratosis and squamous cell skin cancer, their presence or lack thereof could prove important in indicating the onset of cancer. Additionally, because PPARG has been linked to other cancers such as colon and prostate cancer, its expression in the early stages of skin cancer could also serve as a potential indicator for the onset of other types of cancers.

Therefore, we analyzed the expression of the PPARG protein in tissue samples that had mutations in the p53 gene and had variable Aurora Kinase A gene expression.

The p53 gene is responsible for tumor suppression. P53 functions as a transcription factor that encourages formation of the p21 protein which, in turn, prevents a cell division stimulating protein (cdk2) from allowing cells to continue division. A loss of function (LOF) mutation in the p53 gene prevents the gene from suppressing cell mitosis since p21 is no longer produced. When combined with other factors, this mutation could be responsible for the formation of cancerous tumors.[1] A gain of function (GOF) mutation in the p53 gene prevents tumor suppression, thereby encouraging carcinogenic growth. A GOF mutation in the p53 gene has been found to be more detrimental to cells than a LOF mutation because it significantly increases the possibility of metastatic cancer.[6]

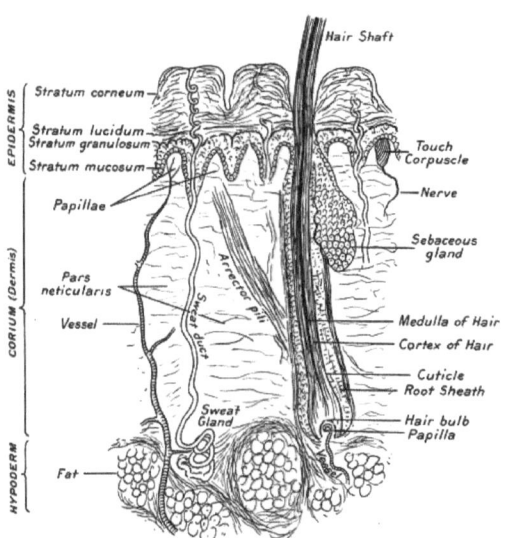

Figure 1: A diagram depicting the layers of the skin. Squamous cell skin carcinoma occurs in the epidermis, specifically the stratus corneum.[3]

The enzyme Aurora Kinase A is involved in healthy cell mitosis. Mutations in the Aurora A protein lead to mitotic difficulties in which chromosome misalignment or cytokinesis fails to occur. Studies have shown that in multiple cancer tumor cells, Aurora A is over-expressed, suggesting it might contribute to tumorigenesis. The p53 gene communicates with Aurora Kinase A and inhibits development of cells. Aurora A attaches a phosphate group to the p53 gene, which in turn degrades the protein. Inhibitors of Aurora Kinase A previously used in cell samples to treat cancer have shown positive results, suggesting that the over-expression of Aurora A could serve as a predisposition for cancer.[2] A lack of Aurora Kinase A will also prevent healthy cell division because it no longer serves as a check in the mitotic process. Therefore, when a lack of Aurora Kinase A is combined with a lack of a tumor suppressor (such as p53), tumors are more likely to develop.[8]

The effects of these genes on cancer have been more widely studied, whereas little is known about the expression of the PPARG gene on squamous cell carcinoma. PPARG is known to suppress the growth and progression of tumors in other cancers. We looked at the PPARG protein in different hyperplastic skin tissue samples. The expression of PPARG in skin cells has been measured at the mRNA level in tissues with LOF and GOF mutations of the p53 gene and in cells that either contain or lack Aurora Kinase A (**Fig. 2**). It was shown that within the mRNA level, the highest expression of PPARG was found in epithelial cells containing a LOF mutation in the p53 gene with Aurora Kinase A present. GOF mutations that contained and lacked Aurora Kinase A had the second highest expression. LOF mutations that lacked Aurora Kinase A had low expression of PPARG at the mRNA level. In this research, we examined whether the expression of the PPARG gene at the protein level is similar to its expression at the mRNA level within hyperplastic skin cells.[6]

For our research, we used immunohistochemistry (IHC) assays to study skin tissue from the backs of mice in the hyperplastic, precancerous stage of squamous cell skin cancer. We characterized the expression of the PPARG protein in tissues containing the Aurora Kinase A protein with GOF and LOF mutations in the p53 gene. We then characterized PPARG in tissues with the same mutations, but without the presence of Aurora Kinase A. We compared data collected in cancer tissue to normal skin tissue. By doing so, we were able to examine the specific patterns of expression of PPARG in hyperplastic skin which may serve as an indicator of the onset of squamous cell skin cancer.

METHODS

Our research involved the characterization of PPARG in four tissue sample types. Paraffin embedded tissue samples from the backs of mice were cut and provided to us by Dr. Enrique Torchia, a co-professor and researcher at the University of Colorado Denver. We looked at four sets of adult mouse tissues per trial; two had Aurora Kinase A present, and of those two, one had LOF and one GOF mutation in the p53 gene. The other two tissues did not have Aurora Kinase A present, but still had LOF and GOF mutation in the p53 gene. Our negative control was non-cancerous tissue from the backs of mice and our positive control was the ABCA12 mutant mice model. This model is homozygous for a mutation that models hyperplastic

squamous cell skin carcinoma. Tissue samples for both the positive and negative control came from embryonic mouse tissue.

We tested two different IHC protocols, one using a mouse antibody and one using a rabbit antibody. During pre-trials, we determined the Mouse on Mouse (M.O.M) protocol was more effective in staining the PPARG protein than the protocol with the rabbit antibodies, as it resulted in a clear stain with less background.

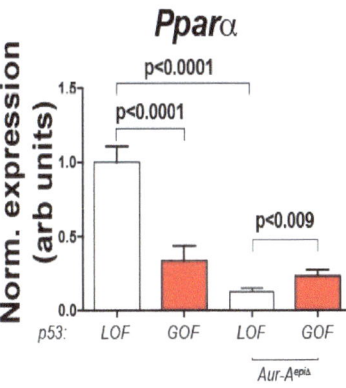

Figure 2: A graph depicting results from previous research done with the PPARG gene at mRNA level. The level of mRNA expression is based on an arbitrary pre-developed scale (y-axis). The first two tissue samples contained the Aurora Kinase A protein and had either a LOF or GOF mutation in the p53 gene. The other two samples did not have Aurora Kinase A but had either a LOF or GOF mutation in the p53 gene.[7]

We characterized the expression of PPARG in these tissue samples using an IHC protocol which involved four main steps: deparaffinization, antigen retrieval, staining, and counter-staining.

Deparaffinization and Antigen Retrieval

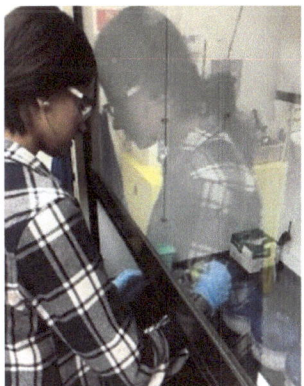

Picture 1: Kumar moves slides through xylene and ethanol gradient for deparaffinization.

To begin, we removed the paraffin in which the tissue samples were embedded by washing them with graded ethanols, including a 100%, 95%, and 70% solution, and xylene (**Pic. 1**). The samples were kept in distilled water to maintain moisture and limit nonspecific antigen-antibody reactions during staining.

After deparaffinization, we used heat induced epitope retrieval to expose specific antigen sites. We made a 20x citrate buffer solution with distilled water. We then placed the samples in a conical vial filled with the citrate buffer and put them in a pressure cooker for antigen retrieval. The slides were washed in a 10x tris buffered saline solution (TBS) and dried before a hydrophobic barrier was drawn around each sample with a wax pencil (**Pic. 2**).

Rabbit Antibody Protocol

After antigen retrieval, we washed the tissue samples in the 10x TBS solution. We then soaked the slides in a 10% normal goat serum blocking solution for two hours at room temperature. Then, we applied our primary antibody

(PPARG rabbit anti-mouse) which was diluted 1:250 to the tissue samples, and incubated overnight at 4°C.

After incubation, the tissue samples were rinsed again in the TBS solution and then the slides were placed into a 0.3% hydrogen peroxide and TBS solution for 15 minutes. Afterwards, we applied the secondary antibody, which was conjugated with the enzyme HRP, to the tissue samples which were incubated at room temperature for one hour. After incubation, we rinsed the tissues with the TBS solution three times for five minutes each.

Mouse Antibody Protocol

After antigen retrieval on these samples as well, the sections were then incubated at room temperature in an incubation chamber with M.O.M Mouse Ig blocking reagent, washed twice with the TBS solution for five minutes and then incubated with the M.O.M diluent at room temperature for one hour.

The primary antibody (PPARG mouse anti-mouse) was prepared at a ratio of 1:500 using the M.O.M diluent, and was then applied to the slides (**Pic. 3**). The biotinylated secondary antibody (anti-mouse IGG) was applied for 30 minutes and then the slides were washed with the TBS solution before staining with diaminobenzidine (DAB).

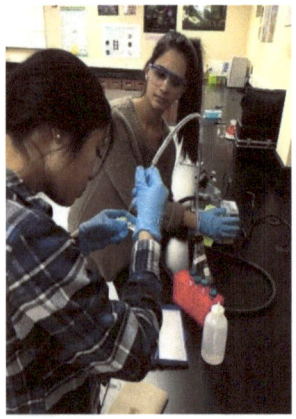
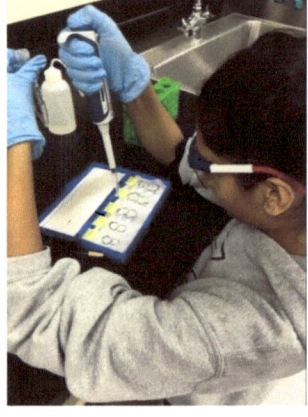

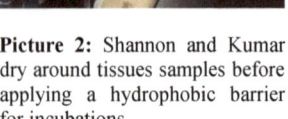

Picture 2: Shannon and Kumar dry around tissues samples before applying a hydrophobic barrier for incubations.

Picture 3: Sharma pipetting the primary antibody onto the tissue sections in the humidity chamber.

DAB Staining and Counterstaining

We placed 10% DAB stain on each sample after performing either the mouse or rabbit antibody protocols, for an incubation time that allowed for adequate coloration on the positive control samples but limited background on the negative control samples. The samples were incubated in DAB stain between one and ten minutes in order to optimize stain but limit background staining. The samples were then counterstained with 100% hematoxylin before being dehydrated and mounted. Sections were mounted with Permount and were then allowed to dry for 4-16 hours before imaging under the microscope. We stained 20 slides, each containing between two and four tissue sections. For each set of slides, we tested one section with the secondary antibody alone (no primary control) and one with both the primary and secondary antibody. This was done to control against nonspecific binding, as well as to optimize the DAB incubation time.

After finishing the experimental trials, we took the stained slides up to our mentor's lab in order to view the results under a 400x microscope. In our positive controls, we expected to see immense brown coloration, as it would show the presence of PPARG in the precancerous model samples and support our hypothesis that a certain expression of PPARG could be an indicator of the onset of squamous cell carcinoma. In our negative controls, we expected to see minimal brown staining present in the epidermis, as there were no mutations in those controls. We also looked for background within the tissue that may suggest nonspecific binding.

RESULTS

When we ran our immunohistochemistry protocols, we analyzed two different antibodies, the mouse antibody and the rabbit antibody. The mouse antibody was then used to stain our experimental slides and control tissue samples. Our initial tests compared the antibodies to determine which produced a better stain of PPARG while limiting background noise. Background noise is any nonspecific binding caused by either the primary or secondary antibody. To analyze the background noise of the primary antibody, we compared staining on the negative antibody to staining on the positive antibody. To analyze background noise of the secondary antibody, we stained all experimental samples with only the secondary antibody.

Picture 4 shows sections from multiple DAB stained slides. The epidermal layer and a portion of the dermal layer are visible, and staining in both the epidermal and dermal layer are seen in all the samples pictured. The background in all samples is high, as many unexpected areas in the dermal layer have staining present.

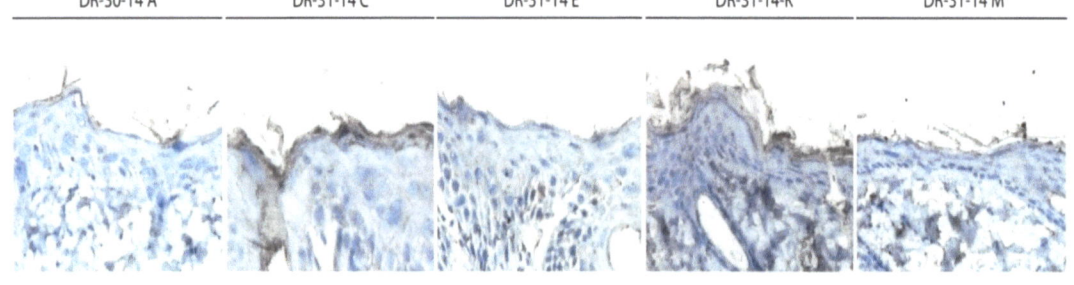

DR-30-14 A   DR-31-14 C   DR-31-14 E   DR-31-14-K   DR-31-14 M

Picture 4: Epidermal layer of our experimental trials stained with DAB. DR-30-14 A has the Aur-A deletion with GOF p53. DR-31-14 C and E have the Aur-A deletion with LOF p53. DR-31-14 K and M have the LOF p53.

Picture 5 shows that both antibodies had significantly stronger DAB staining in the epidermal layer of the positive control (labeled MUT) than in the negative control (labeled WT), as indicated by the increase in brown coloration. The negative control, which was normal embryonic mouse skin tissue for both antibodies, has minimal brown coloration in the epidermis, as expected. The positive control, which was the ABCA12 mutant squamous cell skin carcinoma model, had brown coloration concentrated in only the epidermis, as expected. The mouse antibody, however, reflected a stronger presence of DAB coloration and less background throughout the control tissue samples.

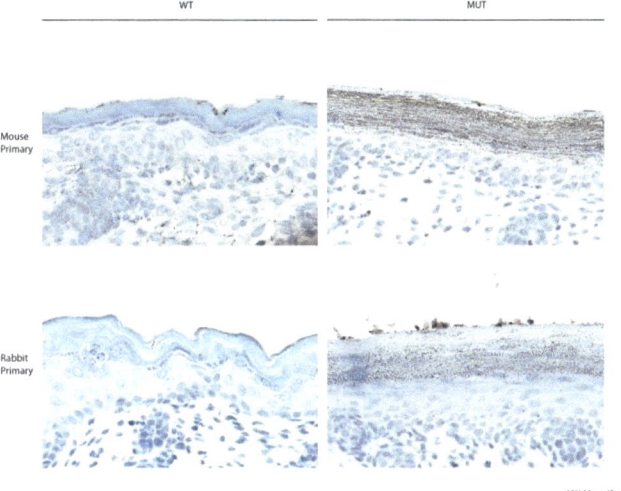

Picture 5: PPARG staining (brown) between the rabbit and mouse antibodies in our positive mutant control and negative wild type control.

Picture 6 shows brown coloration present in all three samples shown. The staining is primarily in the epidermal layer. In the positive samples without any primary antibody, brown coloration present in the epidermal layer shows that background staining occurred from the secondary antibody.

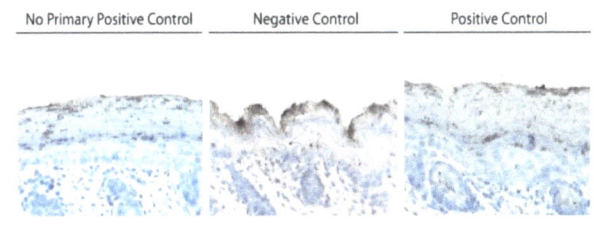

Picture 6: Second staining of control slides with the mouse antibody and an additional no primary positive control.

DISCUSSION

Squamous cell skin carcinoma affects the epidermal layer of the skin and the mechanisms behind its development are primarily unknown. The purpose of our research was to examine the expression of PPARG in precancerous squamous cell skin carcinoma tissue samples and compare the expression to that present in normal skin in order to determine whether the presence of PPARG can be used as an indicator for the onset of skin cancer. Previous research conducted indicated differences in the expression of PPARG at the mRNA level between the experimental groups, suggesting that a difference may also exist at the protein level. We hypothesized that since PPARG is a tumor suppressor, expression will increase in precancerous skin in order to slow the progression of cancer.

After experimentation, however, significant background noise was seen in the tissue samples, which prevented us from fully analyzing and interpreting the data. In our control samples, the negative control had significant brown coloration present. Since our negative control was normal skin tissue that should have had very little PPARG present, this indicated that the antibody was targeting a protein other than PPARG (**Pic. 6**). This differed from our first round of staining, in which the negative wild type control samples had very little brown coloration present in comparison to the positive mutant control (**Pic. 5**).

Unexpectedly high levels of DAB staining were also present within the experimental samples due to nonspecific antibody binding. We expected to see significantly more PPARG present in the epidermal layers, as was evident in the control slides and pre-trials; however, when the adult tissue experimental slides were stained, very little difference was seen between the epidermal layer and rest of the tissue. In addition, the experimental slides stained without a primary antibody showed there was still brown coloration present, indicating the secondary antibody was binding non-specifically (**Pic. 6**).

In the control slides taken from embryonic tissue, PPARG was concentrated in the epidermis, liver, and eye; this was expected because these organs have high fat content and PPARG has function in adipose tissue. Very little staining was seen throughout the rest of the tissue. In the adult experimental samples, PPARG was present at high levels throughout the tissue sections, which could be explained by higher fat content in adult mice. These higher levels of staining throughout the tissue, however, could also be due to nonspecific binding.

Throughout our research, several sources of error impacted our results. The mouse antibody worked very well on both the positive and negative control samples which were embryonic tissues; however, when we attempted to stain our experimental trials with the same antibody, the staining was not as clear and had a high amount of background noise preventing us from being able to quantify the presence of PPARG in these tissues. In addition, we conducted our initial tests without a secondary-only control, so were unable to determine whether the background staining present was caused by the secondary or primary antibody. The initial tests indicated very limited background; however, after staining another round of positive and negative control embryonic tissue samples, each with the secondary antibody only, it was clear that the secondary antibody was targeting other proteins, even if minimally. The no primary antibody positive control sample had some brown coloration present in the epidermal layer, showing that the secondary antibody may have contributed to the brown coloration seen in samples with both antibodies, causing the background to be too high to ascertain the presence of PPARG **(Pic. 6).**

To further this research, a western blot could be conducted using the same antibodies used in the IHC protocol to determine the exact proteins the antibodies were targeting and which proteins were causing the background noise. Furthermore, future researchers could use a different antibody to minimize background staining. The antibody we used was the M.O.M antibody and overall was not optimal when detecting our specific protein.

ACKNOWLEDGMENTS

We would like to especially thank Dr. Enrique Torchia with University of Colorado Anschutz Medical Campus and Gates Center for Regenerative Medicine and Michael Ferreyros, BSc with University of Colorado Anschutz Medical Campus for guiding and supporting our research as well as providing tissue samples, antibodies, and many other materials we needed throughout the project. We would also like to thank them for their help with our data analysis and tissue imaging. We would like to thank our donors, Sudha Sharma, Diane Shannon, and Mohit Kumar, for funding our research and making it possible. We also are thankful to Vector Labs for donating the M.O.M. protocol kit and DAB staining kit essential for our research. We also thank Tom Dillon, with Community College of Aurora, for feedback on our experimental design and Matt Bernstein, with ThermoFisher, for providing instruction needed for our project as well as for donating western blotting supplies for our lab. We thank Susanne Petri and Nikki Dobos, teachers at RCHS, for supervising our research and sharing their lab space with us. We are very thankful to Wendy Lerolland for providing us with editing assistance and support. We thank David Ferguson for help with our distillation process and providing technical support with our methods. We thank Bryan Winkelman for designing our website, guiding us through our blog posts, and providing necessary resources throughout this year. We express gratitude to Amy England and Jim McClurg for filming and editing our video. Finally we thank Rock Canyon High School and DCSD for providing lab equipment and supplies for us to conduct our research.

REFERENCES

1. Bethesda. (1998). The p53 tumor suppressor protein. *Genes and Disease*. NCBI. Retrieved 2016, September 25 from U.S. National Library of Medicine Website. [Web]
2. Fu, J., Bian, M., Jiang, Q., & Zhang, C. (2007). Roles of Aurora Kinases in Mitosis and Tumorigenesis. *Molecular Cancer Research*, 5(1), 1-10. DOI:10.1158/1541-7786.mcr-06-0208
3. Glands [Digital image]. (2017). Retrieved 2017, April 20. [Web]
4. MacNeal, R. J. Structure and Function of the Skin. (2016). *Merck Sharp & Dohme Corp.* Retrieved 2016, September 23. [Web]
5. Squamous Cell Carcinoma. *Skin Cancer Foundation.* (2016). Retrieved 2016, September 23 from Skin Cancer Foundation Website. [Web]
6. Torchia, E. (2016). Personal interview.
7. Torchia, E. [A graph showing the expression of the PPARG protein at the mRNA level throughout various trials.] Retrieved 2016, August 31.[Web]
8. Yang, H., Burke, T., Dempsey, J., Diaz, B., Collins, E., Toth, J., . . . Ye, X. (2005). Mitotic requirement for Aurora A Kinase is bypassed in the absence of Aurora B Kinase [Abstract]. *FEBS Letters, 579*(16), 3385-3391. DOI:10.1016/j.febslet.2005.04.080

ABOUT THE AUTHORS

Pictured: (from left to right) Mallika Kumar, Dr. Enrique Torchia (mentor), Sriya Sharma, Waverly Shannon, and Mr. Michael Ferreyros, B.Sc. (mentor)

We learned many valuable lessons throughout this research. We learned that many times, although the results of an experiment may not be as expected, this does not reduce the value of that experiment, as it may lead to further research. We also learned to work resiliently and diligently. Our research required us to follow protocol strictly and required us to perform many tasks at once, teaching us the importance of being knowledgeable about the protocol before performing steps to ensure it runs smoothly.

The lab skills we gained throughout our project are extremely valuable especially as we look into pursuing scientific fields of study in college. We learned how to work in a lab safely and to follow protocol. We learned many relevant skills for not only our project but also for working in any lab, including how to make sterile agar plates and how to perform dilutions. Throughout our research, we learned how to use lab equipment that gave us the experience needed to obtain internships in our future.

The three of us hope to pursue careers in the scientific field and will be studying either biology or chemistry based majors. We are all interested in the medical field and are excited for our upcoming years.

Assaying the effects of Cannabidiol oil and *Hericium erinaceus* extract on the rates of paralysis of *Caenorhabditis elegans* induced with beta-amyloid peptide modeling Alzheimer's disease

A. E. England, A. R. Lepard, H. J. Philip, H. N. Swamberger, and S. L. Fordham.
Department of Science, Principles of Experimental Design in Biotechnology, Rock Canyon High School, Highlands Ranch, Colorado, USA

Alzheimer's is a chronic disease that affects millions of people all over the world. The disease causes brain tissue and function to slowly deteriorate. Those affected often experience memory loss, speech impairment, behavioral changes, and difficulty performing simple tasks. The purpose of this investigation is to test cannabidiol (CBD) and *Hericium erinaceus* to reduce the rate of paralysis caused by buildup of β-amyloid peptide modeling Alzheimer's disease (AD). Cannabidiol (CBD), a compound found in *Cannabis sativa*, and *Hericium erinaceus*, a medicinal mushroom, have both been recognized as having potentially therapeutic effects for AD specifically.[11] We used CL2006 *Caenorhabditis elegans*, an AD model organism, to test the effects of CBD oil and *Hericium erinaceus* extract on the rate of paralysis. We hypothesized that we would observe a reduction in the rate of paralysis in the *C. elegans* exposed to these treatments. Our data analysis showed that *C. elegans* exposed to CBD and *H. erinaceus* showed a statistically significant reduction in the rate of paralysis compared to the controls. These results are a promising first step in identifying compounds that could have beneficial effects on slowing the progression of AD.

Approximately 5.4 million Americans currently live with Alzheimer's disease (AD) with one out of nine seniors above the age of 65 diagnosed with AD.[1] After diagnosis, a patient is likely to live an additional 8-10 years. In the United States, AD is the sixth leading cause of death.[17] Symptoms of AD include lack of drive, memory loss, loss of coordination, difficulty speaking, delusions, and pain. There is currently no cure and the available treatments tend to have harsh side effects. AD is caused by a buildup of plaques and tangles in the brain which originate in the hippocampus and spread throughout the cortex.[17] The plaques are clusters of β-amyloid peptides that block cell signaling. The tangles are made up of the protein tau, which restricts essential nutrients from moving throughout the brain. It is unknown what causes the plaque buildup and tangles in AD. The clusters of plaques and tangles also trigger an immune reaction, resulting in inflammation of the tissue in the brain. In addition, these plaques and tangles cause neurological cell degeneration and tissue loss in the brain.[3]

In this investigation, we used *Caenorhabditis elegans* modeling AD **(Pic. 1)** to test whether the rate of paralysis could be reduced with exposure to different treatments. *C. elegans* are microscopic nematodes that share fundamental biological characteristics with humans and have a lifespan ranging from 12-18 days. The CL2006 *C. elegans* strain is genetically modified to have the gene that produces β-amyloid peptides (derived from human amyloid precursor protein cDNA), which serves as a model of similar plaque buildup that takes place in humans with AD. The *C. elegans* become paralyzed and subsequently die due to the beta-amyloid (Aβ) toxicity **(Pic. 1)**. The CL2006 strain was created by Dr. Christopher Link of University of Colorado,

Boulder in the Institute for Behavioral Genetics. Dr. Link found a positive correlation between the amount of Aβ plaque and rate of paralysis in these *C. elegans*.[7]

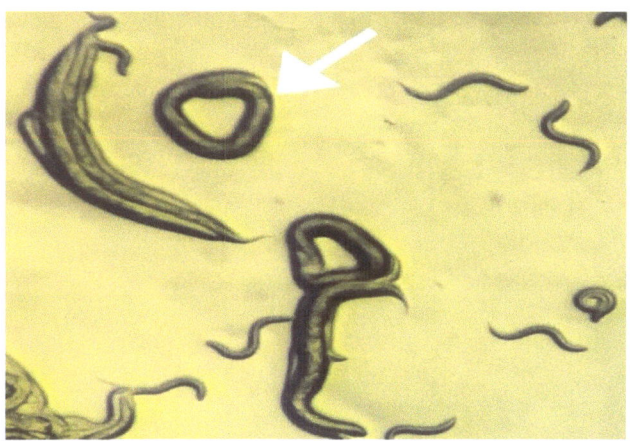

Picture 1: A photo of the CL2006 under a microscope. In this photo you can see different life stages of the nematodes. You can also see the rolling phenotype (arrow) compared to the normal S shaped *C. elegans*.

Previous studies using this strain of *C. elegans* have shown that exposure to coffee throughout the entire lifespan slowed the rate of paralysis and plaque buildup in the AD model *C. elegans*. It was later determined that the coffee triggered the antioxidant pathway which slowed the plaque buildup and subsequent rate of paralysis.[7] Research has found that compounds high in antioxidants trigger a pathway that slows down neurological damage associated with AD.[4] Oxidative stress has also been identified as a significant factor in AD, which would explain why triggering the antioxidant pathway reduces the amount of Aβ plaque buildup in AD.[14]

We identified two compounds with antioxidant properties that may also be able to mitigate the paralysis in the CL2006 *C. elegans*. We measured how exposure to Cannabidiol oil and *Hericium erinaceus*, the lion's mane mushroom, extracts throughout the entire lifespan affected the rate of paralysis in the *C. elegans*.

According to the research cited in a government-issued patent, cannabinoids have been found to be neuro-protectants because they preserve neuronal tissue and prevent damage that happens over time.[10] This could help limit the damage caused by neurodegenerative diseases, such as Alzheimer's and Parkinson's.[8] Many patients diagnosed with AD have anecdotally claimed that cannabis alleviates their symptoms of agitation, aggression, apathy, and delusion; but minimal scientific research has been conducted to support these claims.[16] Cannabis is comprised of 85 known compounds making it difficult to know which compounds, if any, have a positive impact on the symptoms of AD. One of the major compounds is CBD. In-vitro lab studies show CBD as a possible treatment for AD because of its neuroprotective properties and its role in reducing oxidative stress, apoptosis, and inflammation.[14,5] Rats have also been used as model organisms for research with CBD and AD with results showing that CBD reduced neuroinflammation.[9] However, doctors believe that not enough scientific research has been conducted to recommend CBD to be used as treatment for AD.[2]

We used the Charlotte's Web Brand Everyday Advanced Dietary Supplement CBD oil for this experiment (**Pic. 2a**). This brand of CBD oil is confirmed by the State Department of Colorado to have less than 0.3% tetrahydrocannabinol (THC).[18] CBD by itself does not have psychoactive properties, unlike the other well-known cannabinoid, THC. Limiting the THC, allowed us to attribute any changes in the *C. elegans* rate of paralysis specifically to the CBD. Charlotte's Web CBD contains 50 mg of hemp extract per mL with an olive oil base.

Picture 2: (**a**)This is a photo of the CBD we are using in our experiment. It was purchased from CW Hemp and contains a high amount of CBD. (**b**) This is a picture of one of nine jars of dried *H. erinaceus* given to us by Mr. Ward.

The second treatment we tested was *H. erinaceus,* an edible mushroom that carries no psychoactive properties (**Pic. 2b**). *H. erinaceus* has been consumed in China and Japan for thousands of years for both culinary and medicinal purposes. The mushroom can be found in the North American, Asian, and European regions. Previous studies have demonstrated that extracts of *H. erinaceus* have potential therapeutic effects due to its antioxidant, anticancer, and anti-inflammatory properties as well as its ability to stimulate the synthesis of nerve growth factors (NGFs).[6] NGFs help prevent apoptosis of nerve cells and neuronal death which is high in AD. This demonstrates the possibility of *H. erinaceus* to be effective at slowing the rate paralysis in the CL2006 *C. elegans*.[13]

Currently there is a growing need for treatment and preventatives for AD. Both CBD and *H. erinaceus* are potential oral supplements that show potential to have therapeutic effects for AD because of their antioxidant properties that could help slow the β-amyloid peptide buildup.

METHODS

Treatments of Cannabidiol oil and *H. erinaceus* were tested on the genetically modified *C. elegans* strain CL2006. This strain has been engineered to produce Aβ plaques which we targeted with our treatments. The paralysis of the *C. elegans* associated with these plaques allowed us to test the effectiveness of these treatments on the disease model.

To examine the effects that the CBD oil and *H. erinaceus* extract had on the rate of paralysis in the AD modeling *C. elegans*, we spread the extracts onto the 100mm compartmented Y-petri plates containing nematode growth media (NGM) agar. The tri-welled plates, purchased from IPM Scientific, allowed us to run three trials at once. To avoid cross contamination, each section of the petri plate received the same treatment. We had five treatments in this experiment: CBD oil extract, *H. erinaceus* extract, both CBD oil and *H. erinaceus* extracts, olive oil, and NGM media only. We used extra virgin olive oil as a control because the CBD oil we used contained an olive oil base, and olive oil also has antioxidant properties. This control allowed us to compare changes in the rate of paralysis in the plates containing CBD oil to the olive oil control. The plate with no treatment also served as a second baseline to compare the rate of paralysis to.

Solution Preparation

We used an ethanol extraction protocol to extract the heri-cenones and erinacines from the *H. erinaceus* (**Pic. 3**). After freezing the mushroom, we used a food processor to grind the *H. erinaceus* into a fine powder. We then created a 7g/ml *H. erinaceus* solution using 95% ethanol.

Picture 3: The centrifuged extract used for the plates

After mixing the solution, we incubated it at room temperature for 30 minutes and centrifuged the sample to extract the supernatant. We then heated the extraction until it boiled to evaporate the majority of the ethanol.

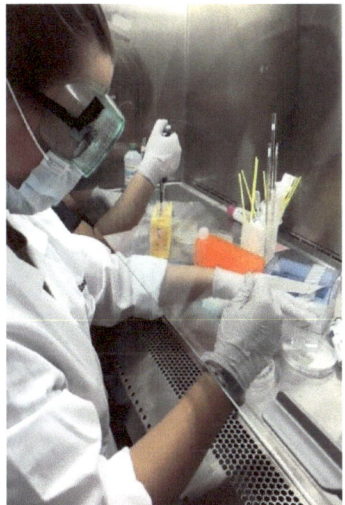

Picture 4: England seeding the trial plates with OP50.

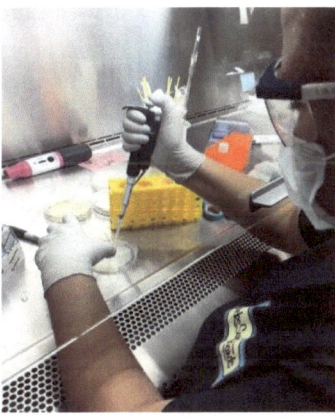

Picture 5: A photo of the CBD plate being prepped by Swamberger in the biological safety cabinet.

Plate Preparation

The plates were pre-seeded with the OP50 strain of *Escherichia coli* obtained from the Dolan DNA Learning Center in Cold Springs Harbor, New York (**Pic. 4**). Then, 30µl of each treatment was added to each well on each plate. We performed this step in the biological safety cabinet to avoid contamination (**Pic. 5**). A sterile loop was used to spread the treatment on the surface of the agar to ensure the treatments were spread evenly on the plate. The CBD oil and olive oil treatments tended to bead up on the surface instead of absorbing into the agar.

After all five plates were prepared, each with a different treatment, the plates were dried in the biological safety cabinet overnight at room temperature (**Pic. 6**). This was especially important for the *H. erinaceus* plates to evaporate the ethanol off of the plates.

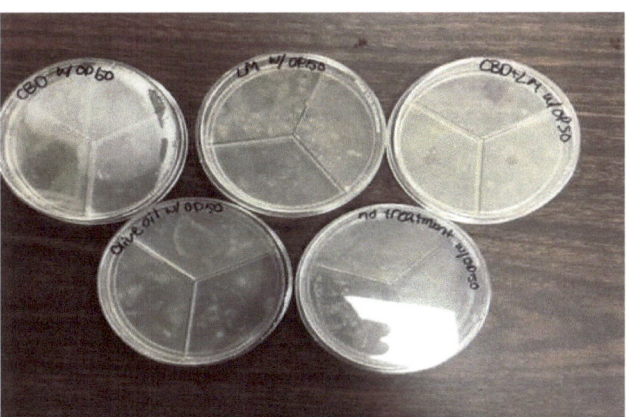

Picture 6: All five plates seeded with OP50 and treatments.

Experimental Methods

We picked approximately 5-7 fertile adult *C. elegans* onto each section of the petri plates. We let them lay eggs for 6-7 hours and then picked the adult worms off the plate creating a synchronized population. After one day, when the eggs matured to the L1 stage, we documented the rate of paralysis for all *C. elegans* in each

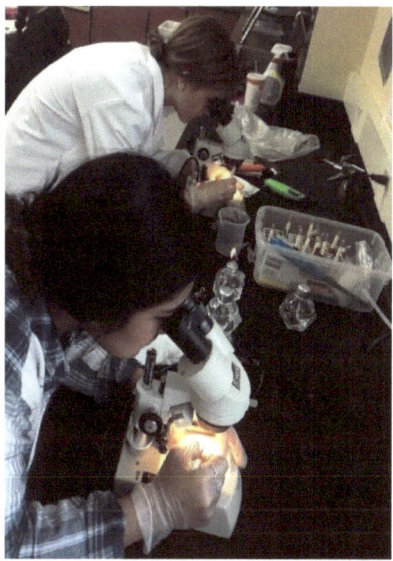

Picture 7: Lepard (top) and Philip (bottom) picking.

well of each plate and picked off dead *C. elegans* as they were identified (**Pic. 7**). This was repeated every other day for 10 days.

The rate of paralysis of each *C. elegans* was documented using the following categories: not affected, semi-paralyzed, fully paralyzed, and dead. A non-affected *C. elegans* was defined as swimming or moving normally; a semi-paralyzed *C. elegans* was defined as only moving the head and tail; and a fully paralyzed *C. elegans* had no movement except for the head. A dead *C. elegans* showed absolutely no movement.

RESULTS

Every other day for 10 days we recorded the stage of paralysis of CL2006 *C. elegans* on every plate. We conducted several statistical tests to compare the rate of paralysis between our treatments. Throughout our trials, it was evident that each treatment had changed the rate of paralysis, but in order to determine if our results were significant we performed multiple 2 proportion z-tests. We used these z-tests to calculate a z-scores which we then converted into p-values. P-values were used to make sure that our results occurred due to our treatments, not random chance.

CBD showed a significant reduction in the total paralysis compared to its controls (**Graph 1**). By day ten, *C. elegans* on the CBD plate had a total paralysis of 44.28% whereas *C. elegans* on the no treatment plate had a total paralysis of 81.42% and *C. elegans* in olive oil had a total paralysis of 68.75%. The most significant result that we calculated from this treatment occurred on day four. The CBD vs. no treatment plate had a z-score of 4.885, which showed us that the difference was due to the treatment and not chance. The rate of paralysis for the CBD treatment is significantly lower than either of its controls starting at approximately day six.

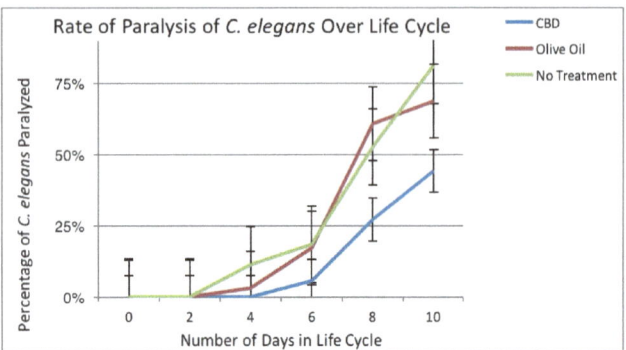

Graph 1: The rate of paralysis of CBD compared to its controls.

H. erinaceus also resulted in a statistically significant lower total paralysis compared to the *C. elegans* receiving no treatment **(Graph 2)**. The *H. erinaceus* exposed *C. elegans* only had 52.23% of the total population paralyzed on day ten as compared to the 81.42% of the previously mentioned control. On day two, *H. erinaceus* had a z-score of -4.0362 and on day four it was 4.3608. This shows a statistically significant difference between the treatment and control. On day six the rate of paralysis for the no treatment control spikes where the lion's mane rate of paralysis is much slower in comparison.

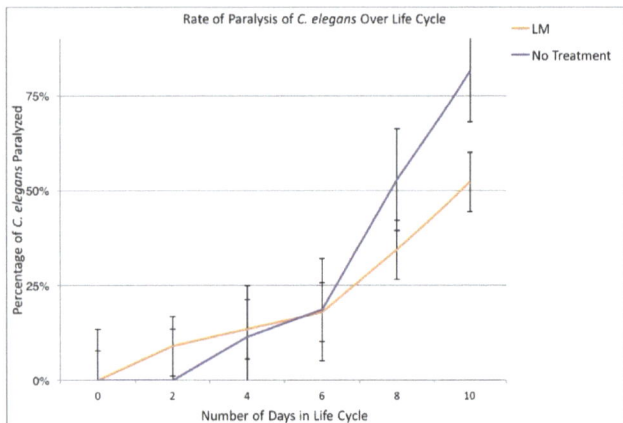

Graph 2: The rate of paralysis of lion's mane mushroom (LM or *H. erinaceus*) compared to the control.

Between the CBD, *H. erinaceus*, and a combination of both treatments, shows no significant difference in the total population paralyzed in the *C. elegans* **(Graph 3)**. 37.14% of the total population of *C. elegans* living on the plate with both CBD and *H. erinaceus* had paralyzed on day ten which is lower than either treatment alone. This indicated that together the two treatments may have an additional effect on lowering the overall paralysis. The rate of paralysis for all three treatments followed very similar trends.

We compared the treatments in this experiment to each of their controls as well as to each other. Using a 99% confidence interval, the largest measurable difference that we observed was on day four between CBD and the no treatment plate. Based on our p-value the difference that we observed in the amount of *C. elegans* paralyzed between the two plates would occur 1 in 1,934,235 times if it were to occur by chance. Other days that had high z score values were days two and four of *H. erinaceus* vs. no treatment.

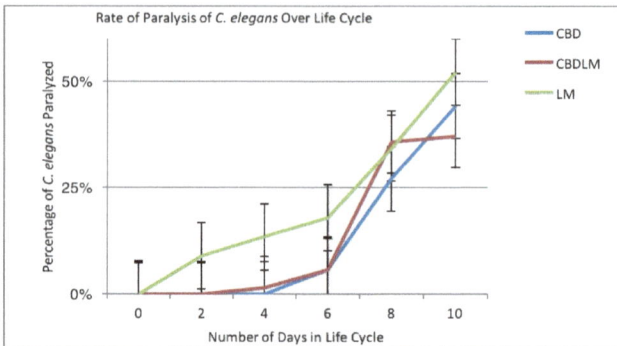

Graph 3: This line graph shows the difference in the rates of paralysis for three different treatments, CBD, CBD/*H. erinaceus*, and LM. Also, lion's mane mushroom (LM) is *H. erinaceus*.

DISCUSSION

AD is a fatal disease affecting millions worldwide. Recently there has been increasing demand for natural treatments for AD. CBD and *H. erinaceus* are strong candidates that show preventative properties against AD. In this research, we used *C. elegans* modeling AD to test the effects of two treatments, CBD oil and *H. erinaceus* extract, on the rate of paralysis. Our original hypothesis was that we would observe a statistically significant reduction in the rate of paralysis in the *C. elegans* exposed to CBD and *H. erinaceus*. In addition to comparing the CBD and *H. erinaceus* to our controls, we also compared the treatments against each other to see which had a greater effect. After analyzing our data, we found a strong trend showing the *C. elegans* exposed to CBD and *H. erinaceus* had a statistically significant lower paralysis rate than those with no treatment. By the end of the ten-day trial, there was a noticeable difference between the CBD, *H. erinaceus*, and the combined plates and the control plate. The total population CBD plate remained below the controls of olive oil and no treatment throughout each day of the experiment **(Graph 2)**.

On day ten, plates with CBD treatments had little progression of paralysis while the olive oil and no treatment plates had a significantly higher amount of *C. elegans* in late stages of paralysis. The percent of worms paralyzed on the last day of observation was only 44.28% on the CBD plate, 52.23% on the *H. erinaceus* plate, 37.14% on the CBD and *H. erinaceus* plate, compared to the controls, which were 68.75% for the olive oil and 81.42% for the no treatment. Near the beginning of the trial, we noticed the plates with *H. erinaceus* had an early progression of paralysis but this slowed after day six. The difference between *H. erinaceus* and the control was insignificant until day 6 when the rate of paralysis of the *H. erinaceus* treated *C. elegans* became noticeably less compared to the rate of paralysis with controls **(Graph 3)**. This trend with slower rate of paralysis is similar for all treatments, indicating all treatments have the same positive effect on slowing the rate of paralysis **(Graph 1)**.

In comparison, similar research has been conducted on CL2006 *C. elegans* modeling AD using coffee. Exposure to coffee, which is high in antioxidants, was found to significantly lower the rate of paralysis of the CL2006 *C. elegans*. On the 5th day of adulthood, they found that

CL2006 exposed to coffee presented 73.6% paralyzed while the control had 100%.[12] It was later discovered that coffee triggers the antioxidant pathway in the *C. elegans*.

There has been a noticeable trend between products with antioxidant properties and the reduction of Aβ. In addition to CBD, *H. erinaceus*, and coffee, products such as dark chocolate, blueberries, and artichokes are high in antioxidants, and may also have protective effects against AD.[15]

Our results could potentially be compromised due to possible sources of error. One of these sources include the concentration of ethanol to *H. erinaceus* used. Although our protocols allowed for the ethanol to evaporate from the *H. erinaceus* plates, we were not able to measure how much, if any, ethanol remained on the plates. The ethanol is toxic to *C. elegans,* so having more than anticipated may reduce the positive impact of the *H. erinaceus*. It is possible that even greater positive results would have been observed without this variable. Another source of error stems from having four researchers collect the paralysis data using the paralysis scale. Although each stage of paralysis was discussed thoroughly, it is probable that one person considered the stage differently than another. Running three trials simultaneously is also a potential source of error in that we didn't test that our results are repeatable when run for a second or third time.

Using antioxidants to slow progression of AD is very promising based on our results. The next step needed to continue our research would be to confirm our results using different model organism. An ideal model organism to use in the next study would be a mammal, either a rat or mouse. This study would help to determine whether these results could be mimicked in other animal studies.

ACKNOWLEDGMENTS
Throughout this research we have had substantial support financially, technically, and scientifically. We would like to thank Dr. Christopher Link from the University Colorado, Boulder for providing us with the CL2006 *C. elegans* strain and giving us useful tips to use throughout our experiment. His past research helped shape our own project and identify the right approach to our methods. We are grateful for Matthew Ward with the Life Science Department Laboratory with the State of Colorado who graciously donated samples of the *H. erinaceus* mushroom and provided us with the extraction protocols. We are also grateful for Allie Kellner, a former RCHS Biotech student, for her guidance with our protocols. We also thank the *Caenorhabditis* Genetics Center at the University of Minnesota for their donation of additional CL2006 *C. elegans*. We appreciate Officer Erik Brown for providing a safe location for the CBD oil. A grant from Groove Automotive allowed us to purchase a microscope attachment which aided in visualization of our results. We would like to thank Merlin Kunnel, Bethany Gallegos, Kini Spence, Don and Joanne Dhein, Tom and Patricia Swamberger, and Cheri and Wilma Gailunas for generously donating money to fund our research. We also are very thankful for Brian Winkleman help with planning our website, blog posts, and with other editorial assistance. We would also like to thank Jim McClurg for filming and editing the video for our website. David Ferguson, a Rock Canyon High School Chemistry Teacher, also helped us with the mushroom ethanol extraction. We appreciate Dr. Jason Dunkle for assistance with our data analysis and Wendy Lerolland for an abundant amount of editorial assistance and support. We would also like to thank Tom Dillon for his support and feedback with our experimental design. Again, we appreciate Susanne Petri and Nikki Dobos for sharing laboratory space with us and Rock Canyon High School and Douglas County School District for providing the lab space and the equipment used throughout experiment. We are grateful for all of the support.

REFERENCES
1. Alzheimer's Disease Facts and Figures. (2016). Alzheimer's Association. Retrieved 2016, September 25. [Web]
2. Belendiuk, K. A., Baldini, L. L., & Bonn-Miller, M. O. (2015). Narrative review of the safety and efficacy of marijuana for the treatment of commonly state-approved medical and psychiatric disorders. *Addiction Science & Clinical Practice*, 10(1). DOI:10.1186/s13722-015-0032-7
3. Brain Plaques and Tangles. (2000). Alzheimer's Association. Retrieved September 22, 2016. [Web]
4. Calabrese, V., Bates, T.E. & Stella. (2000). NO synthase and NO-dependent signal pathways in brain aging and neurodegenerative disorders: The role of oxidant/antioxidant balance. A.M.G. *Neurochem Res*. 25: 1315. DOI:10.1023/A:1007604414773
5. Campbell, V. A., & Gowran, A. (2009). Alzheimer's disease; taking the edge off with cannabinoids. *British Journal of Pharmacology*, 152(5), 655-662. DOI:10.1038/sj.bjp.0707446
6. Chang HC, Yang H-L, & Pan J-H. (2016). *Hericium erinaceus* inhibits TNF-α-induced angiogenesis and ROS generation through suppression of MMP-9/NF-κB signaling and activation of Nrf2-mediated antioxidant genes in human EA.hy926 endothelial cells. *Oxidative Medicine and Cellular Longevity*, 2016,8257238. DOI:10.115 5/2016/8- 257238.
7. Dostal, V. & Link, C. D. (2010). Assaying β-amyloid toxicity using a transgenic *C. elegans* model. *J. Vis. Exp.* (44), e2252. DOI:10.3791/-2252
8. Eshhar E. (1995). Neuroprotective and Antioxidant Activities of HU-211, A novel NMDA receptor antagonist. *European Journal of Pharmacology*, 283,19-29.
9. Esposito, G., Scuderi, C., Valenza, M., Togna, G. I., Latina, V., Filippis, D., & Steardo, L. (2011). Cannabidiol reduces Aβ-Induced neuroinflammation and promotes hippocampal neurogenesis through PPARγ involvement. *PLoS ONE*, 6(12). DOI:10.1371/journal.pone.-002- 8668
10. Hampson, A.J., Axelrod J., & Grimaldi, M. (2003). U.S. Patent No 663,0507. Washington, DC: US Patent and Trademark Office.
11. Iuvone, T., Esposito, G., Esposito, R., Santamaria, R., Rosa, M. D., & Izzo, A. A. (2004). Neuroprotective effect of cannabidiol, a non-psychoactive component from *Cannabis sativa*, on beta-amyloid-induced toxicity in PC12 cells. *Journal of Neurochemistry* . 89(1), 134-141. DOI:10.1111/j.14714159.2003.02327.x
12. Kellner, A., Galyon, B., & Fordham, S. (2016). Assaying the effects of Namzaric and coffee on paralysis in beta-amyloid peptide Alzheimer's disease model *Caenorhabditis elegans*, strain CL2006. *Research in Biotechnology*, 1(1), 7-13.
13. Ma B., Shen J., Yu H., Ruan Y., Wu T. & Zhao X. (2010). Hericenones and erinacines: stimulators of nerve growth factor (NGF) biosynthesis in *Hericium erinaceus*, Mycology, 1:2, 92-98. DOI:10.1080/21501201003- 735556
14. Markesbery, W. R. (1997). Oxidative stress hypothesis in Alzheimer's disease. *Free Radical Biology and Medicine*, 23(1), 134-147.
15. Mathis, C. E. (2005). 20 Common Foods with the Most Antioxidants. WedMD. Retrieved 2017, April 12. [Web]

16. Shelef A., Barak Y., Berger U., Paleacu D., Tadger S., Plopsky I., & Baruch Y. (2016). Safety and efficacy of medical cannabis oil for behavioral and psychological symptoms of dementia: an-open Label, add-on, pilot study. *Journal of Alzheimer's Disease JAD* 51.1 (2016): 15-19. DOI:10.3233/JAD-150915

17. Shenk, D. (2010). Understand Alzheimer's Disease in 3 Minutes. Retrieved 2016, September 11. [Web]

18. The World's Most Trusted Hemp Extract™. (2016). Stanley Brother Enterprises. Retrieved 2016, September 16. [Web]

ABOUT THE AUTHORS

Pictured: From left to right, Haley Swamberger, Ariel Lepard, Hannah Philip, and Amy England. Mentor Allie Kellner (not pictured).

As a group we have gained knowledge and experience incomparable to other high school students. The Experimental Design in Research Biotechnology opened up doors to us that we never knew existed. We got hands on experience on how research is planned, conducted, and analyzed. The research we conducted was also comparable to groundbreaking studies that are being conducted on medical marijuana. No high schooler in the state or district has done research quite like ours, and we are proud of overcoming the obstacles we faced with this project. Throughout the year we learned how to deal with adversity inside and outside of our research and learned how crucial communication is to a successful research project. We all hope to take what we have gained to help us further our education and our careers on day.

Biotech has always been a class I (Amy) wanted to take and I was lucky enough to have been selected for Research in Biotechnology, which opened doors and my eyes to the in's and outs of how real graduate level research works. Conducting this research was very interesting and a huge part of my senior year as it pertains to me because my great grandmother passed from Alzheimer's. Because of this course, I have decided to study Biomedical sciences and become a Pediatric Physician's Assistant.

Science was always something I (Ariel) excelled at throughout my education, but I never thought I would see myself doing it for my career until I took Biotech. This career field is promising and something I have a deep passion for. I have learned so much from skills, teamwork, and just conducting and completing research. I am looking to study biomedical sciences or microbiology.

I (Hannah) have always known I wanted to pursue a science based career, and the biotechnology program has sparked my interest. Through the class I have been able to learn the skillsets any researcher would need as well as comprehend scientific writing and data at a new level. In addition to the knowledge I gained from this course, I was given several opportunities to network and acquaint myself with professionals in the field. I plan focus on the business aspects of a biotechnology or pharmaceutical company.

I've (Haley) never been one to particularly love science until I took the Introduction to Biotechnology class. After taking the first year course I knew that I wanted to push myself and discover what else this field has to offer. The Research in Biotechnology class did just that. I discovered a door to a world that I did not know existed. I plan to study Finance and International Relations and plan to be a financial advisor for a professional sports team.

Developing a microfluidic attachment for foldscopes allowing for blood analysis

K. S. Franklin, A. P. Jerath, C. J. Weintraub, and S. L. Fordham
Department of Science, Principles of Experimental Design in Biotechnology, Rock Canyon High School, Highlands Ranch, Colorado, USA

A foldscope is a paper pocket microscope that can be easily transported due to its durability and ease of assembly. Developed at Stanford University by Prakash Labs, these paper microscopes allow for field microscopy to take place; ranging from educational to medical diagnostics, the foldscope has revolutionized microscopy. In order to fully utilize the medical diagnostic potential of foldscopes, we designed and tested several microfluidic device that works jointly with foldscope in order to help field practitioners diagnose blood diseases such as malaria and African sleeping sickness in environments lacking advanced medical diagnostic equipment. We tested three prototypes: a nail polish based design, plastic capillary tubes, and glass capillary tubes. We found that the stretched glass capillary tube prototype was the most successful, due to its clarity and relative ease of creation. Additionally, the design is affordable and most medical teams would have these materials already. We also found that the nail polish design was far too variable to reproduce and was not durable enough. In addition, the plastic capillary tube, while cheap and easily designed, had such a low level of clarity that the design would not be successful. If this design is expanded upon, it has potential practical use in the field or as an educational tool.

In 2011, Manu Prakash and his team identified the need for a durable, lightweight, and inexpensive microscope for easy use in the field, where such equipment isn't readily available such as in developing countries. In these locations often lack electricity as well as access to sophisticated medical equipment. A foldscope is a 6 cm by 17.5 cm paper based microscope that can be transported in a person's pocket **(Pic. 1)**. It costs less than a dollar, and it requires no electricity to operate.[5]

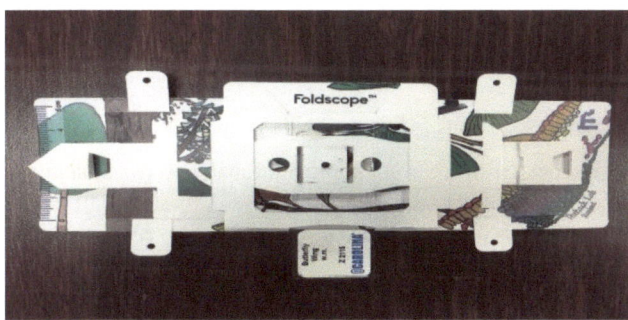

Picture 1: Image of our completed foldscope with slide.

The foldscope comes in a kit sent from Prakash Labs, a subsidiary of Stanford University. The kit contains the unassembled foldscope, a lens, a small lithium ion battery, and an instruction sheet. The foldscope is made through simple folding and assembly. Foldscopes have many attachments that make them easier to use, including magnets for attaching to a phone to make projection and taking photos possible, a portable light, and paper slides. One advantage of the paper-based microscopes is that they are incredibly durable. They are able to withstand falling three stories and are also waterproof.[7] These advantages are critical for field use microscopy as normal microscopes are fragile, hard to transport and require electricity.

Currently, foldscopes are being used in the field as an educational tool.[6] The beta program for the foldscopes began in 2014 and has nearly 10,000 adopters worldwide. Preliminary work has begun on using foldscopes for diagnosing diseases.[6] In 2014, specific clinical models of the foldscope were provided to clinical field teams in India and the Philippines for analyzing bacteria and other microorganisms. Other uses of the foldscope include looking at the microbiomes of water filled potholes.[6] Due to the lack of infrastructure in developing countries, namely roads and electricity, field teams are looking for cheap and efficient ways to diagnose illnesses without the high cost of transporting and purchasing expensive, fragile equipment. The foldscope aids in the diagnostics of multiple diseases.

The true advantages of the foldscope appear when comparing them to normal microscopes. Normal microscopes are fragile and require detailed and safe transportation, which is not feasible in much of the world, African Highway is the main road that serves the area. This road has many missing parts, and some areas that are impossible to pass after rain storms; in some areas the road does not exist.[8] These bumpy road conditions are impractical for transporting a microscope safely. This makes the durability of the foldscope important, as it can be easily transported in the roughest conditions without risk of breaking.

The magnification of foldscopes is comparable to normal microscopes. The foldscopes have lens magnifications ranging from 140X to 2,000X, allowing their use to vary from an educational tool to a useful field microscope. The most common lens magnification for a foldscope is the low magnification lens at 140X. However, multiple attachments allow for higher magnification. In comparison, the maximum magnification reached by the light microscopes at RCHS is 100X, which is less than the lowest magnification offered by foldscopes. For our purposes, we used a lens that allows for 400X to 600X magnification in order to match the magnification required to observe the red blood cells and protists that cause African sleeping sickness.[2]

Visualization of blood allows for the diagnosis of many different diseases. For example, through the analysis of the morphology of a red blood cell, doctors can diagnose malaria, sickle cell anemia, and various other blood borne illnesses, while the number of red blood cells in a sample can reveal malnutrition, dehydration, and pulmonary fibrosis.[1] The blood plasma can show parasites within the blood that cause African sleeping sickness.[2] On the other hand, identification of malaria requires a Giemsa stain of the blood to highlight the specific malformations on the red blood cells that characterize malaria **(Pic. 2)**.[3] Last year, 216,000,000 cases of malaria resulted in approximately 438,000 deaths. In the African region, only 65% of cases were identified, meaning that 35% of cases were undiagnosed.[9] Those living with undiagnosed malaria do not receive proper medical care and treatment and can spread the disease to others through contact with infected mosquitoes and blood transfusions. Effective blood diagnostics impacts the treatment of many deadly diseases. A microfluidic attachment for the foldscope would allow for the diseases to be diagnosed in a wide range of environments. Additionally, using the phone attachment would allow the user to count the blood cells as a sample of the whole, as well as observe a larger volume of blood than a blood smear on a slide.

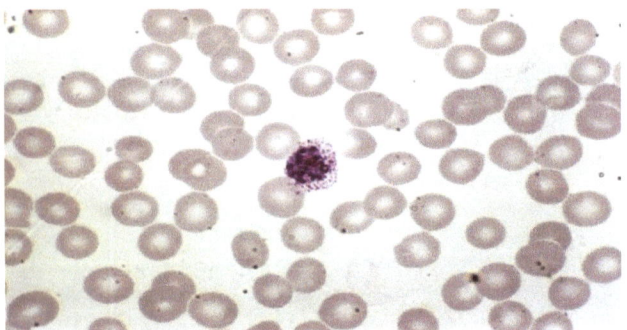

Picture 2: Image of a blood cell with malaria in the middle of many normal red blood cells.

Our goal was to create an efficient, cost effective and practical attachment for the foldscope allowing for small amounts of blood to pass through the magnifying lens portion on the foldscope.

METHODS/ RESULTS
Foldscope Assembly
After we received our foldscope from our mentor Damon Tighe with BioRad, we assembled it to begin our research. The foldscope gifted to us had the lowest magnification available, 140X, compared to their highest magnification of 2000X. Fully assembling the foldscope took approximately 30 minutes the first time.

Nail Polish Prototype

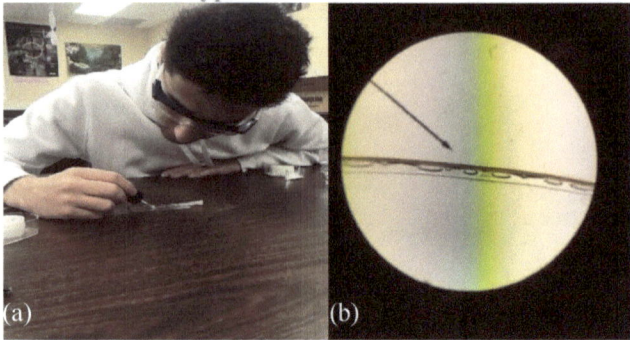

Picture 3: (a) A. Jerath applying nail polish to the slide over a hair to create the nail polish channel. **(b)** An early nail polish channel without the hair removed.

To further our mentor's research, we began by developing a nail polish prototype by coating hair with nail polish **(Pic. 3a)**. Using human hair at the start of prototyping we coated the hair in transparent nail polish in several thin layers to reduce bubbling. The goal was to remove the hair after drying leaving a channel on the glass microscope slide shown in **(Pic. 3b)**. We also attempted to create a larger channel using a strand of .013 in fishing wire. The fishing wire pulled up the nail polish when we attempted to remove it from the slide, making this option not viable. As a solution to, we also tried horse hair due to its thickness which worked better.

Nail Polish Prototype

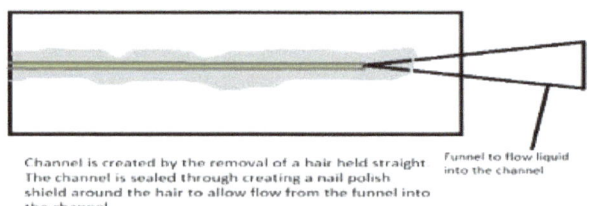

Channel is created by the removal of a hair held straight. The channel is sealed through creating a nail polish shield around the hair to allow flow from the funnel into the channel.

Funnel to flow liquid into the channel

Picture 4: A diagram of the nail polish channel prototype.

In order to move liquid into the channel, we attached a micropipette tip to the slide and coated this with nail polish to create a seamless channel that would prevent leaks **(Pic. 4)**.

This prototype is cheap to produce and easily transportable but was relatively challenging to assemble and had variable success rates. The main problem with the design was that the application of the nail polish created bubbles and holes that caused the channel to leak. As well, the design process took a large amount of time. Compared to the other porotypes the nail polish channel was extremely difficult to use and not practical for use with foldscope.

Plastic Capillary Tube Prototype
For our second prototype, we decided to use plastic capillary tubes to see whether they would be a better option than the nail polish channel in terms of clarity, durability, and ease of use.

Picture 5: Franklin testing the collection bag and plastic capillary tube design.

Capillary tubes are inexpensive and readily available; and move blood and other liquids using capillary action. Capillary tubes are a great option for a prototype because they are currently being used by medical field teams around the world. We began our design using plastic capillary tubes that were provided to us by our fellow classmate Sahana Narayan. We developed a method of collecting the liquid so that it could be pushed and pulled through the capillary tube using a pipette, or a syringe, multiple times and quickly disposed of using a common medical latex glove (**Pic. 5**). Cutting off the finger of a glove and using a rubber band to secure it to the capillary tube, created a closed system without leaks that could quickly be exchanged for a clean finger and new blood sample (**Pic. 6**).

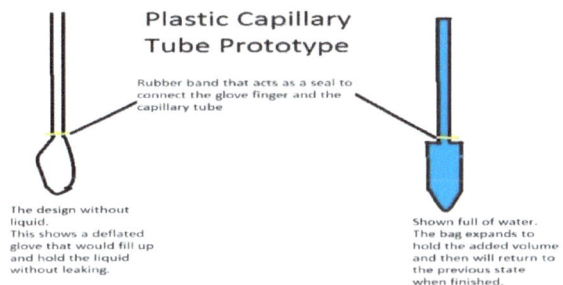

Picture 6: A diagram of the plastic capillary tube prototype.

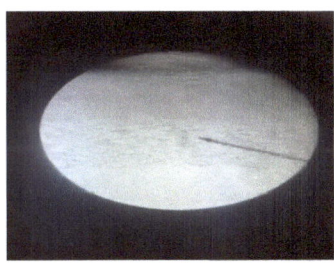

Picture 7: The plastic capillary tube containing green algae cells underneath a microscope.

Examination of the plastic capillary tubes under the microscope and our own plastic material was too coarse to see through, and greatly obscured clarity (**Pic. 7**). To combat this issue, we tried to cut the tube in half to improve the clarity and reduce the working distance. This proved unsuccessful because the capillary tubes were too small to easily be cut.

Glass Capillary Tube Prototype

For the third prototype we used glass capillary tubes, due to their improved clarity. After consulting our mentor, we began to stretch the tubes using fire to melt the glass in order to restrict the amount of cells that could pass through the tube to the point that individual cells could be observed. After heating the glass we then pulled it apart gently

(**Pic.8**). Due to the low temperature of the flame, heat resistant gloves were unnecessary as the heat transfer in glass was not enough to burn at the edges of the capillary tubes. This method worked well and was easy to replicate. We were concerned that the burning would scar or fog the glass but this was not the case (**Pic. 8**).

Picture 8: (Left) Weintraub stretching the glass capillary tube using a picking flame. (Right) An image of the algae under the microscope, due to the lack of magnification the cells are hard to see.

The glass capillary compared to the plastic had drastically less visual obscurities, but, due to the fragility of glass, the tubes would crack and break under unwanted pressure. To combat this issue, we taped the capillary tube to a glass slide to stabilize it. The glass slide actually enhanced the image of the liquid inside the capillary tube, because the light from the Foldscope LED wasn't refracted off the cylindrical surface. We tried to use the same collection bag design from the plastic capillary tubes, but it did not work. The glass capillary tube cut a hole through the collection bag. In the end, this was unnecessary because we just used the capillary action of the tube instead of a pumping mechanism. We accomplished this by using a piece of cloth or paper towel to pull the liquid through the tube.

Picture 9: A diagram of the glass capillary tube prototype.

Testing of Prototypes

After the conclusion of the design phases of the experiment, we moved into testing. We had each of the students in our class test the plastic and glass capillary tubes using the algae instead of blood and rank them on a scale of 1 to 5 with respect to their clarity. On this scale, a score of 1 represented the algae barely being visible, a 3 meant the student could make out the algae within the channel but not visualize it in detail, and a 5 indicated perfect clarity with sharp, clear images.

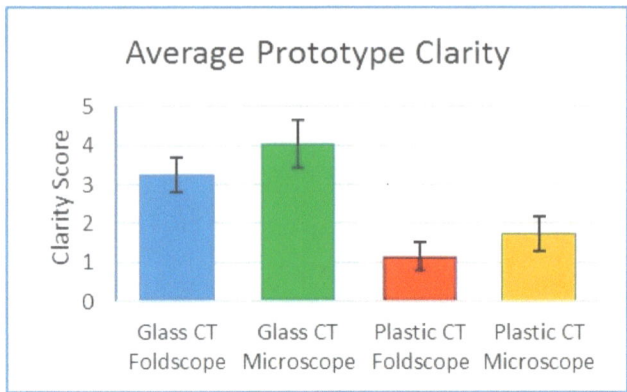

Average Prototype Clarity

Graph 1: A graph of the average student's rating of clarity on the 1 to 5 scale with error bars detailing standard deviation.

We tested only the plastic and glass capillary tubes because the nail polish channel was deemed unsuccessful. In the testing, we had each subject look at the glass and plastic capillary tubes under a microscope, then look at the glass capillary tube prototype under the foldscope that we had assembled. Graph one shows the average student rating of clarity for each design, with the error bars showing the standard deviation. No prototype met the needed visibility level that would be sufficient for clinical use in the field, however, we were unable to test these prototypes using the higher magnification lens that foldscope has been recently created.

DISCUSSION

With the foldscope and a means to observe blood in a field setting, clinical teams could have a cheaper and more efficient way to diagnose blood based illnesses. In this project we attempted to design a modification to foldscope technology that would allow its use to expand beyond the primary educational goal, giving it functionality with field diagnostic work. By adding a microfluidic attachment our designs were focused on adhering to the spirit of the foldscope in being easy to assemble and inexpensive to produce.

The nail polish channel had success early on, but we discovered that the channel would often have random bubbles causing leaks. As well, the channel only lasted a short time before collapsing and took extensive practice to perfect. Due to this, we decided that the design was impractical and too time consuming to be useful.

The plastic capillary tube used in conjunction with a syringe and medical glove was effective at creating a leak-free system that was inexpensive. It met all of the criteria except the clarity of this design made it unusable, as you could only see the rough outlines of the algae, if you could see them at all. Visualization of the blood cells is crucial for field diagnostics.

Our last prototype was the glass capillary tube. It had the highest clarity of all three prototypes. The stretched glass was not scarred from the flame and was able to restrict the flow to individual cells through the area of the eye piece. Due to the nature of capillary tubes, capillary action allowed the blood to flow through the tube without any external force. The only thing needed to perform capillary

action was a paper towel or cloth to slowly wick the blood through the channel. When compared to the normal microscope, the glass capillary tube under the foldscope dropped by nearly a point on the clarity scale. We attribute this to the early model foldscope that we had. Two years ago, Foldscope began a beta program to generate interest and find potential issues; we obtained one of these foldscopes, and conducted our research using this foldscope. In October 2016, Foldscope launched a Kickstarter to fund their global release and will We believe that the new foldscope will be easier to use making the design more apt for medical use in the field.

Our research had some limitations resulting from the time frame and our laboratory's biosafety level. We started our project as foldscope was undergoing a global launch, missing the release date by a few months. Due to this, we were unable to obtain a foldscope with the higher magnification that would enable us to see the cells clearly through the foldscope. In future studies, our glass capillary tube prototype needs to be tested with the newer model at a higher magnification. In addition, the prototype needs to be tested using real blood. Our lab is a BSL-1 lab and we cannot have real blood in the lab. For future research with this microfluidic attachment, researchers need to work in a BSL-2 or 3 lab that would allow for testing the concept with blood and even blood that shows evidence of disease.

Our testing process has found that the glass capillary tubes are the most successful prototype due to their widespread accessibility, low cost, and high clarity. By packaging our directions with Foldscope Instruments, we believe our design will aid in the diagnosis of blood diseases and will have a positive impact in the world. We hope the lowered clarity observed with the foldscope is negligible in the future with the launch of an improved model.

ACKNOWLEDGMENTS

We would like to thank our mentor Damon Tighe from Bio-Rad for giving us guidance, support, and encouragement that allowed our creativity and innovative spirit to flourish. This project would not have been possible without him. We would like to thank Cindy Davis and Kristin Franklin for funding our project. We would also like to thank Sahana Narayan for providing us with the capillary tubes to us through her volunteer work at Sky Ridge hospital's NICU. We would also like to thank all of the people who made this project possible. Thank you to Nikki Dobos for sharing her lab space, David Ferguson for providing materials and ideas, and Bryan Winkelman for helping to publish the journal, setting up the website, and guiding us through our blog posts. We would like to thank Wendy Lerolland for her editorial assistance and support she has given us throughout this entire process. Thank you to our college advisor Tom Dillon for providing helpful advice and guidance through the earliest parts of our project. We would also like to thank Rock Canyon High School and Douglas County School District for supporting our research by providing laboratory space and equipment needed to perform our research.

REFERENCES

1. Cafasso, J. (2015). Red Blood Cell Count (RBC). Retrieved on 2016, September 14. [Web]
2. Center for Disease Control. (2012). African Sleeping Sickness Diagnosis. Retrieved 2016, September, 26. [Web]
3. Center for Disease Control. (2016). Malaria Diagnosis. Retrieved 2016, September 26. [Web]
4. Cybulski, J. S., Clements, J., & Prakash, M. (2014). Foldscope: Origami-Based Paper Microscope. *PLoS ONE, 9*(6). DOI:10.1371/journal.pone.0098781.
5. Foldscope FAQ. (2015). Foldscope. Retrieved 2016, September 26. [Web]
6. Prakash, M. (2015). Microcosmos. Foldscope. Retrieved 2016, October 16. [Web]
7. Prakash, M. (2012). TED Talks: A 50-cent microscope that folds like origami. Retrieved 2016, September, 14.
8. Review of the Implementation Status of the Trans African Highways and the Missing Links. United Nations Economic Commission for Africa. (2003). SWECO International. Retrieved 2016, September 25. [Web]
9. World Malaria Report 2015. (2015). World Health Organization. Retrieved 2016, September 14. [Web]

ABOUT THE AUTHORS

Pictured: From left to right, Colby Weintraub, Kyle Franklin, Damon Tighe, Ansh Jerath.

Over the course of this class, we have not only developed a prototype to solve a real problem within the world, we have gained a new understanding of what it means to be an engineer. We learned how to work and communicate as a team and how to successfully fail and try again. Additionally, we have learned how to innovate, and that this process is continuous. There are always aspects you can improve upon within your design, no part is ever perfect. This class has taught us how to be resilient, and taught us how to deal with stress in a constructive manner. In the past year this class has brought us together, not just our group, but the entire class. This class was not only unique in content, but unique in terms of the dynamics we had with one another. Unlike other classes in the school, this class motivates you to pursue your dreams and make a difference.

I (Colby) have always wanted to be an entrepreneur. This class helped me realize the importance of keeping a schedule and taught me how to make people believe in a collective goal. I (Kyle) have wanted to be an engineer for many years, and this class taught me how to collectively come up with innovative ideas. I (Ansh) have always loved making things, physical or electronic. Looking back at what we accomplished in this class, my most important takeaway is learning what it is like to work in a lab, and work under pressure.

Agrobacterium tumefaciens mediated transfection of various floral species with the *AtPCS1* gene

N. F. Watervoort, M. A. Gilbert, and S. L. Fordham
Department of Science, Principles of Experimental Design in Biotechnology, Rock Canyon High School, Highlands Ranch, Colorado, USA

Heavy and transition metal pollution is a growing problem in the United States and abroad due to increasing mining demands, electronics recycling, industrial processes, and natural phenomena. While many methods exist to combat this issue, growing support has been building for phytoremediation, or the use of plants to positively impact the environment. This experiment was designed to test the ability of *Agrobacterium tumefaciens* to transform different flowering plants with the pART27-AtPCS1 plasmid using floral dip, seed infection, and co-cultivation protocols. The *AtPCS1* gene, from *Arabidopsis thaliana*, encodes a protein that sequesters metal ions from the soil into the biomass.[4] Plants with this gene are able to uptake metal contamination from the soil; however, these protocols are largely untested with the *AtPCS1* gene. We identified the capability of all three methods to integrate *AtPCS1* and found the highest success with seed infection. Co-cultivation failed to produce results and floral dip has not resulted in seeds for testing in time for this publication. Using the seed infection protocol, we created two recombinant plants containing the *AtPCS1* gene that could potentially remediate environmental damage due to metal contamination. *AtPCS1* is a relatively untested gene, and this experiment will serve to build a base for future experimentation.

Metal pollution in Colorado is a widespread issue deriving from various natural and human processes. Within Colorado, over a thousand inactive mining operations are causing environmental damage due to metal pollution. Mines in Leadville, Colorado have been leaching metals into the groundwater and soil for years.[7,14] Additionally, electronic waste disposal facilities in southeast China have been similarly linked to increased levels of metals in soils in the surrounding area.[6] While no study has been conducted relating to electronic disposal in Colorado specifically, similar facilities exist in Grand Junction and Denver with potential to generate issues in the future.[5]

Metal contamination has widespread effects on the microbial processes and groundwater purity in any environment.[8] This can lead to adverse human and wildlife health effects. A potential solution is phytoremediation, or the use of plants to identify and fix environmental hazards. Its attractiveness derives from its use of natural processes, which eliminates the need for harsh chemicals that could cause additional harm to the environment if mishandled. Previous research has shown that plants can be engineered to withstand high metal concentrations by sequestering the metals into their biomass. After growth has ceased, the plants can then be removed from the tainted site and easily disposed of.[4] One such method of disposal is pyrolysis, or the burning of biomass in an anaerobic environment to separate it into its constituent parts.

Engineering plants with a phytochelatin (PC) gene allows them to be useful for phytoremediation as phytochelatins sequester or chelate metal ions into the plant that expresses them. These proteins are found in many plants and are a part of their environmental metal stress responses.[4] Dr. Joseph Jez has pioneered much of this research, finding that

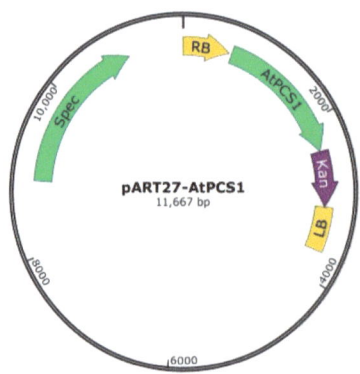

Figure 1: Shown here is the pART27-AtPCS1 vector, which was used as a vehicle for *AtPCS1* transformation. It containes *Spec* and *Kan* markers for cell culture and *AtPCS1*. All of these features are identified as rrrows on the map. This plasmid was ligated by Dr. Joseph Jez prior to this experiment in the pART27 [plasmid backbone

when the transformed plants are removed, metal contamination in the surrounding soil can be removed with them.[3] This quality gives phytochelatins a unique potential for metal phytoremediation. The phytochelatin used in this experiment is the *Arabidopsis thaliana* Phytochelatin Synthetase 1 (*AtPCS1*) gene. AtPCS1 was used due to its higher affinity for metal sequestration compared to other PCs, such as TaPCS. Plants that integrate this gene can remediate contaminated soils faster and more effectively than through other methods.[4]

Many methods exist currently to transform plant species, such as biolistics (gene guns), and *Agrobacterium* mediated transformation. *Agrobacterium* is often used due to its ability to integrate a large sequence of interest reliably at a low cost and decreased copy count.[9] *Agrobacterium* mediated protocols rely on *Agrobacterium*'s natural capability to integrate a sequence of DNA into a host through the use of a tumor-inducing (Ti) plasmid.[11] A custom Ti plasmid can be used to integrate an engineered sequence of transfer DNA (T-DNA) into the host genome. The custom plasmid we used in our research is pART27-AtPCS1, which contains the *AtPCS1, Kan*, and *Spec* genes, which are all transformed into the plant at once (**Fig. 1**). The *Kan* and *Spec* elements of the plasmid infer antibiotic resistance to kanamycin and spectinomycin to cells that take in the T-DNA. The plasmid was ligated by Dr. Joseph Jez before the onset of the experiment.[3] *Agrobacterium* C58 pART27-AtPCS1::FLAG was utilized as the vector for *AtPCS1*. We utilized seed infection, co-cultivation, and floral dip protocols, and tested for the presence of the *AtPCS1* gene after transfer.

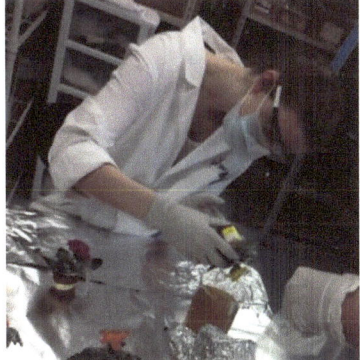

Picture 1: Above, a *Lantana camara* plant is exposed to the floral dip solution of sucrose, Silwet L-77, and *Agrobacterium.*

Seed infection relies on *Agrobacterium* infecting the germ of the seed before any significant growth in order to generate a transformed plant. The seed must be dipped as soon as the seed coat is split by the radical. Seed infection is a less commonly used protocol, because the limited amount of biomass is easily overrun and destroyed by the *Agrobacterium*.

Floral dip is an increasingly popular method for transformation. Buds are submerged in a solution of transformed *Agrobacterium*, a surfactant, and sucrose (**Pic. 1**).[15] *Agrobacterium* infects the bud with the custom plasmid and thus transfers the gene or genes of interest into the forming seeds. If successful, these transformed seeds will retain the gene throughout their lifetime.[16]

Co-cultivation uses the same principle as seed infection, transforming cells from which a plant will grow, but with more biomass. In this protocol, plant tissue or a callus is exposed to *Agrobacterium* to allow for transformation of the tissue. This tissue is then cultured in MS media to create a transformed callus.

METHODS
Three types of tissue (seeds, calli, and buds) were exposed to *Agro-bacterium* to transform them in this experiment.

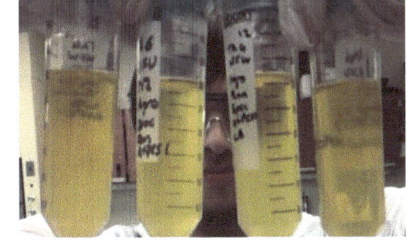

Picture 2: Above are some of the cultures used in this experiment. The culture is composed of LB broth and 50 µg/mL kanamycin and spectinomycin to grow *Agrobacterium* C58 pART27-AtPCS1.

Agrobacterium Solution
Agrobacterium tumefaciens C58 pART27-AtPCS1::FLAG was obtained from

Ashley Sherp, a graduate student in the lab of Dr. Joseph Jez of Washington University in St. Louis. This sample was immediately cultured in LB broth with 50 µg/ml concentrations of kanamycin and spectinomycin.[1] This culture was placed on a shaking incubator at 28 °C and 200 rpm for 32 hours and growth was visually confirmed (**Pic. 2**). This culture was used as our pure *Agrobacterium* solution.

Floral Dip Solution
A floral dip solution composed of *Agrobacterium* C58 pART27-AtPCS1::FLAG, .02% Silwet L-77, and 25% sucrose was also prepared. The *Agrobacterium* contained in the floral dip culture is the same as used in the previously detailed *Agrobacterium* culture.

Seed Infection
Morning glory (*Ipomoea purpurea*), nasturtium (*Tropaeolum majus*), and daisy (*Bellis perennis*) seeds were prepared for germination by lining the bottom of a 100 mm petri plate with absorbent filler paper and placing 40 seeds of each species into separate plates (**Pic. 3**). 2% gibberellic acid was added to the lining of each plate to induce germination. Upon the splitting of the seed coat, 12 seeds were immediately planted and received no treatment. 12

Picture 3: Seen above are the *I. purpurea* seeds used in the seed infection trials. They were used at this state in particular as they just split the testa (seed coat) and would have the highest chance of yielding transformation.

separate seeds were submerged for 3 sec in the *Agrobacterium* culture, then resubmerged for 3 sec in a 100 µg/ml solution of ampicillin (**Pic. 4**). 12 additional seeds were submerged for 3 sec in the floral dip solution and were

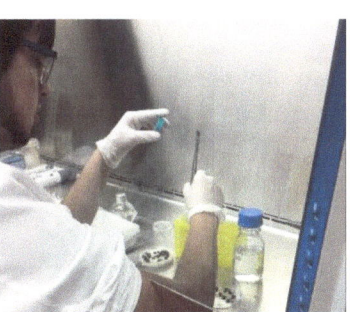

Picture 4: Shown above, Watervoort is performing the process of seed infection with *T. majus* seeds.

then resubmerged for 3s in a 100 µg/ml ampicillin solution.

Every seed was subsequently planted in Black-Gold Gardening Soil and grown indoors on a grow light cart set to a 13 hour light/11 hour dark photoperiod. Every specimen was watered with approximately 7 ml of water daily. When sufficient growth was achieved, two specimens were selected from each experimental group to undergo a DNA extraction.

Verification of Transformation
DNA was extracted from mature tissue of the plants grown

from the seed infection treatment **(Pic. 5)**. A hole-punch sized sample was taken from the leaf of two of the *Agrobacterium* solution trans-formed specimens, two of the floral dip solution transformed specimens, and the non-transformed control. Samples were taken from each species. DNA extraction was performed following a silica resin based protocol outlined in Carolina Biological's DNA Barcode Amplification Kit. 20 µl of extracted DNA was amplified with 2.5 µl of 10 µM T7 promoter primer (3'-TAATAC-GACTCACTATAGGG-5') and 2.5 µl of 10 µM T7 terminator primer (3'-GCTAGTTATTGCTCAGCGG-5') in a PCR tube with a PuReTaq PCR Ready-to-go PCR bead from GE Healthcare. To amplify the DNA, the thermocycler was programmed as follows: The lid temperature of 105°C was constant throughout the program. Initial denaturing occurred at 95°C for 5 min followed by 5 cycles of: 94°C, 55°C, and 72°C for 30s each; 25 cycles of 15s at 94°C, 15s at 55°C, and 30s at 72°C; and 2 cycles of 10s at 92°C, 20s at 55°C, and 90s at 72°C. The samples were then held at 4°C. The PCR products were visualized with a 1.2% Lonza Fast Gel and compared against the pBR322 DNA-BstNI Digest ladder. A band size of approximately 1300 bp was expected.

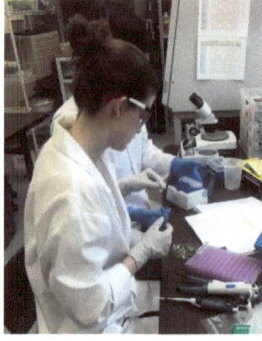

Picture 5: Shown above is the DNA extraction performed on the plant samples. A hole-punch sized sample was used for each extraction.

Two experimental PCR products, one floral dip solution and one *Agrobacterium* solution, along with controls from the daisies and morning glories were sent to GENEWIZ for Sanger sequencing. The nasturtium controls did not grow.

Floral Dip

Floral dip was performed on the flowers of four *Lantana camara* (tickberry) and four *Gerbera jamesonii* (Barberton daisy). A singular flower was selected from each plant, and the rest of the flowers were cut from the plant. The selected flower was submerged up to the stem in the floral dip solution and then gently agitated **(Pic. 6)**. The whole plant was then placed in a plastic bag and kept overnight in the dark. The plants were then removed from the bags and placed on a plant cart with a 13 hour light/11 hour dark photoperiod.

Picture 6: Left: Gilbert adds sucrose to the floral dip solution to improve efficiency of transformation. Right: Gilbert submerges the flower and stem of an *L. camara* in the floral dip mixture. Plastic is visible at the base of the plant, placed there for stability.

Some *Agrobacterium* growth was seen on the stems of the plants after dipping. This infection was treated with several applications of 10 µg/ml ampicillin. Unfortunately, the seeds from these flowers are being produced at a slow rate and have not produce results in time for this publication. Any seeds obtained will be dried before being stored for future researchers to test.

Co-Cultivation

Seven African violet (*Saintpaulia ionantha*) calli were treated by submerging them in the *Agrobacterium* solution followed by a 10 µg/ml ampicillin solution, then placed in African Violet Multiplication Media obtained from Carolina Biological **(Pic. 7)**. A second group of seven calli were submerged in the floral dip solution, followed by the 10 µg/ml ampicillin solution. One callus was not exposed to the *Agrobacterium*, and was kept as a control. Additionally, 3 ml of each *Agrobacterium* solution was placed in an African Violet Multiplication Media vial with no callus to confirm the viability of the *Agrobacterium* in each. All calli were kept

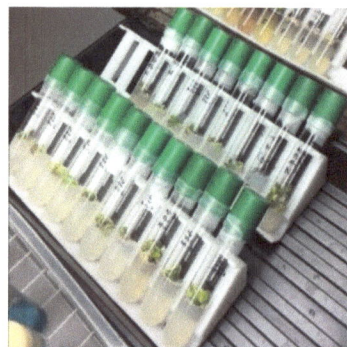

Picture 7: All African Violet calli were placed on the light cart, including both experimental groups, controls, and calli rejected due to preexisting contamination.

under a 13 hours light/11 hour dark photoperiod.

A large amount of microbial growth was observed in every culture vial. 10 µg/ml ampicillin was applied to the contamination in an attempt to kill the excess microbial growth in the culture. All attempts to quell the growth failed, and all calli died in the media, as evidenced by the loss of any green viable tissue.

RESULTS

Three protocols were performed seed infection, floral dip, and co-cultivation in an effort to induce transformation of the AtPCS1 gene into various flora. The seed infection protocol had the greatest success rate. Both *Ipomoea purpurea* (morning glory) and *Tropaeolum majus* (nasturtium) were transformed with the *AtPCS1* gene using the seed dip; however, the *Bellis perennis* (daisy) did not show evidence of transformation **(Pic. 8)**.

Samples that produced a band of approximately 1300 bp suggested transformation.[2] Samples that only produced a band of approximately 20 bp did not appear to be transformed. The *B. perennis*, *T. majus* floral dip solution experimental group, and the control showed only a ~20 bp band, whereas the *I. purpurea* floral dip solution and pure *Agrobacterium* solution groups as well as the *T. majus* pure *Agrobacterium* solution group produced a ~1300 bp band and appear to have successfully transformed. These transformed samples were sent for sequencing, including one floral dip solution transformant and one *Agrobacterium*

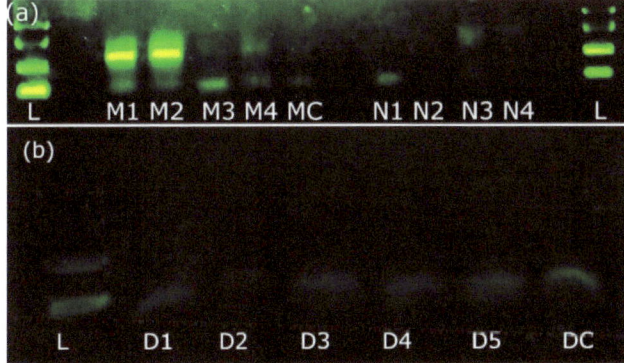

Picture 8: (a) The gel visualizing the transformation of on the *I. purpurea* and *T. minus* samples. M1, M2, N1, and N2 are the seeds dipped in the floral dip solution, M3, M4, N3, and N4 were dipped in pure *Agrobacterium*, and C is control. M and N designate the species, with M being morning glories and N being nasturtiums. **(b)** The gel ran on the *B. perennis* samples with the same numbering scheme as before except with D designating daisy. Transformation was established because of the clear bands created by the experimental morning glory and nasturtium groups. These bands were not seen in the daisy groups which, thus, did not indicate transformation.

Picture 9: Above, the seed infection plants are shown, notably the morning glories (*I. purpurea*) which are seen behind the nasturtiums *(T. majus)*. No daisy (*B. perennis*) survived DNA extraction. The Barberton daisies (*Gerbera jamesonii*) are visible to the right.

Picture 10: (a): A floral dipped experimental group *Lantana camara* plant, this is one of four plants. **(b):** A floral dipped experimental group *Gerbera jamesonii* plant after all but one flower was removed. **(c):** The dead flower of a floral dipped experimental group *Gerbera jamesonii* plant.

background noise, optical bleed through, and low signal intensity.

The floral dipped plants did not produce results in time for this publication. All *L. camara* plants, control and experimental, produced seeds in time for publication but no tests have been performed on the seeds to verify transformation.

The control *G. jamesonii, also* produced numerous seeds; however, none of the experimental treatment for the species has produced seeds (**Pic. 10**). The experimental treatments flowers matured similarly to the control but without producing seeds and eventually died (**Pic. 10**).

The calli exposed to *Agrobacterium* died due to excess microbial growth (**Pic. 11**). The microbes rapidly grew in the agar alongside the calli, overwhelming the calli despite the addition of ampicillin. These samples did not produce results as there was no successful plant growth after transformation. No results could be obtained from these samples.

Picture 11: Shown above are two of the calli after they were dipped. The microbial growth appeared to have halted any further plant growth.

DISCUSSION

This project was intended to determine the effectiveness of three protocols seed infection, co-cultivation, and floral dip on the integration of the *AtPCS1* gene for potential use in future phytoremediation efforts. If transformed, these plants could mitigate the harmful environmental effects of metal contamination in soils worldwide.

AtPCS1 is a relatively untested gene, and this experiment served to test its capability for transformation into *Ipomoea purpurea* (morning glory), *Tropaeolum maju*s (nasturtium), and *Bellis perennis* (daisy) through seed infection; *Lantana camara* (tickberry) and *Gerbera jamesonii* (Barberton daisy) through floral dip; and *Saintpaulia* supports

solution transform-ant from each species with the controls from the *I. purpurea* and *B. perennis*. The *T. majus* controls did not grow and were not sequenced. Unfortunately, the quality scores on all sequences were too low to accurately draw conclusions due to large amounts of

successful transformation occurred with the seed infection protocol with two of the three species - *I. purpurea* and *T. majus*. The observed 1300 bp band that resulted after PCR of this gene region was found in *I. purpurea* and *T. majus* seed infection experimental groups and is similar in bp to the 1300 bp *AtPCS1* gene. The non-transformed experimental groups, *B. perennis*, did not have a band in the 1300 bp region (**Pic. 9**). We were unable to obtain high-quality sequence data of high enough quality to further confirm successful transformation. Therefore, considering that transformation was only confirmed through PCR, conclusions cannot be drawn from our data.

B. perennis is a less domesticated species compared to *I. purpurea* and *T. majus* which are genetic hybrids of multiple species. For this reason, *B. perennis* may have resistance to infection *Agrobacterium* resistance which may have been lost in the other species through cultivation. Alternatively, *B. perennis* has a considerably smaller seed than *T. majus* and *I. purpurea*, making any cells killed by *Agrobacterium* more substantial and more likely to kill the plant.

If confirmed as transformed, these plants may have the ability to uptake metals, like cadmium, from contaminated soil as was demonstrated in yeast cells transformed with this gene in previous studies.[3] If *I. purpurea* (morning glory) was in fact transformed, it should have a potential in phytoremediation due to its rapid growth, large amount of biomass, and relatively extensive root system. These traits allow *I. purpurea* to sequester more metals from deeper in the soil into its biomass. In addition, *I. purpurea*, a climbing vine, is less affected by spatial constraints. *I. purpurea* is also quite hardy and grows in degraded and disturbed sites including those in Colorado.[13] *I. purpurea* should be thoroughly contained if it is ever used in phytoremediation, as it is classified as an invasive species in Arkansas and Arizona. Therefore, *I. purpurea* cannot be used in any phytoremediation efforts.

Tropaeolum majus (nasturtium), if transformed as expected, also has potential in phytoremediation due to its quick growth and ability to spread, approximately 10 feet with a surface area, approximately 3 feet at maturity.[12] Though not native to North America, the species has been introduced to California and several states along the Appalachian Mountains, which are mining centers.[10] *T. majus* is a cold sensitive plant, which makes the plant much less suited to harsh climates in higher altitude regions, like the Rocky Mountains. This makes phytoremediation in Colorado much less probable, but still possible in warmer areas in Colorado.[9]

The mortality rate of the co-cultivation samples was much higher than expected. The microbial growth that overwhelmed the calli was not identified, yet it significantly inhibited plant growth, and the tissue failed to differentiate into shoots and roots.

Floral dip was performed using a floral dip solution. The floral dipped *L. camara* plants have produced seeds that have been dried and stored for future researchers to test. The floral dipped *G. jamesonii* have not yet produced seeds and have dead flowers. The cause of this death is unknown, although the control *G. jamesonii* did produce seeds. As such, the success of this floral dip protocol cannot yet be determined at the time of this publication.

While our initial goal was to transform local Colorado wildflowers, this was not feasible due to the timing of the research and the availability of plant materials. It is important that the transformation of *I. purpurea* and *T. majus* from this experiment is verified by future researchers via DNA sequencing to confirm successful transformation. Researchers should also test the ability of the transformed *I. purpurea* and *T. majus* to uptake metals from the soil and identify where these metals accumulate in the plant. Additionally, researchers could grow many generations of the transformed *I. purpurea* and *T. majus* and determine whether the *AtPCS1* gene is retained across several generations. Wildflowers natural to the Colorado area could also be tested by future researchers with *Agrobacterium* C58 pART27-AtPCS1::FLAG to assess the ability of the plants to be transformed using the seed infection protocol with a floral dip solution.

ACKNOWLEDGMENTS

We are very thankful to have received help from a variety of individuals during this project. We would like to express our immense gratitude to Uma Venkitanarayanan for developing many of our protocols and for providing support and guidance throughout our project. Karin and Frank Watervoort have our incredible thanks for entirely funding our research. Thank you to Dr. Joseph Jez and Ashley Sherp from Washington University in Saint Louis who provided us with *Agrobacterium tumefaciens* C58 pART-27 AtPCS1::FLAG which was integral in our project. Thank you to Wendy Lerolland for providing us invaluable editorial assistance and Bryan Winkelman for setting up our research website, instructing us on our blog posts, and providing continual assistance behind the scenes. Jim McClurg and Amy England also have our thanks for their help with our video. We also would like to thank Matthew Gracey and David Ferguson for providing reagents, supplies, and advice on our project, and Rock Canyon High School and Douglas County School District for providing laboratory space and equipment used in our research.

REFERENCES

1. Antibiotic Concentrations for Bacterial Selection. (n.d.) *Addgene*. Retrieved 2016, September 20. [Web]
2. *Arabidopsis thaliana* phytochelatin synthase 1 (AtPCS1) gene, complete - nucleotide. (n.d.) *GenBank*. Retrieved 2016, September 20. [Web]
3. Cahoon, R. E., Lutke, W. K., Cameron, J. C., Chen, S., Lee, S. G., Rivard, R. S., ... & Jez, J. M. (2015). Adaptive engineering of phytochelatin-based heavy metal tolerance. *Journal of Biological Chemistry*, 290(28), 17321-17330.
4. Cobbett, C. S. (2000). Phytochelatins and their roles in heavy metal detoxification. *Plant Physiology*, 123(3), 825-832.
5. Electronics and computer waste. (n.d.) *Colorado Department of Public Health and Environment*. Retrieved 2016, September 26. [Web]
6. Fu, J., Zhou, Q., Liu, J., Liu, W., Wang, T., Zhang, Q., & Jiang, G. (2008). High levels of heavy metals in rice (Oryzasativa L.) from a typical E-waste recycling area in southeast China and its potential risk to human health.*Chemosphere*, 71(7), 1269-1275.
7. Handy R. M. (2015). Geology, drainage, laws decrease odds of toxic mine spill in Teller, El Paso counties. *The Gazette*. Retrieved 2016, September 25. [Web]
8. Hao, X., Taghavi, S., Xie, P., Orbach, M. J., Alwathnani, H. A., Rensing, C., & Wei, G. (2014). Phytoremediation of heavy and transition metals aided by legume-rhizobia symbiosis. *International Journal of Phytoremediation*, 16(2), 179-202.
9. Piers, K. L., Heath, J. D., Liang, X., Stephens, K. M., & Nester, E. W. (1996). *Agrobacterium tumefaciens*-mediated transformation of yeast. *Proceedings of the National Academy of Sciences*, 93(4), 1613-1618.
10. Plants Profile for *Tropaeolum majus* (nasturtium). (2017). United States Department of Agriculture. Retrieved 2017, April 6. [Web]
11. Pletsch, M., de Araujo, B. S., & Charlwood, B. V. (1999). Novel biotechnological approaches in environmental remediation research. *Biotechnology Advances*, 17(8), 679-687.
12. *Tropaeolum* (group). (n.d.). *Missouri Botanical Garden*. Retrieved April 6, 2017. [Web]
13. White, M. (2017). Invasive plants and weeds of the national forests and grasslands in the southwestern region. 2nd ed. *United States Department of Agriculture*, Retrieved 2017, April 6. [Web]
14. Wood, D. W., Setubal, J. C., Kaul, R., Monks, D. E., Kitajima, J. P., Okura, V. K., ... & Woo, L. (2001). The genome of the natural genetic engineer *Agrobacterium tumefaciens* C58. *Science*, 294(5550), 2317-2323.

15. Zhang, X., Henriques, R., Lin, S. S., Niu, Q. W., & Chua, N. H. (2006). Agrobacterium-mediated transformation of *Arabidopsis thaliana* using the floral dip method. *Nature Protocols*, *1*(2), 641-646.

16. Zupan, J. R., & Zambryski, P. (1995). Transfer of T-DNA from *Agrobacterium* to the plant cell. *Plant Physiology*, *107*(4), 1041.

ABOUT THE AUTHORS

Pictured: (from left to right) Nathan Watervoort, Uma Venkitanarayanan, and Martina Gilbert.

Through the last ten months, we have gained invaluable real-world lab experience and skills. Beyond the basic principles of Introduction to Biotechnology I, we have grown our ability to design and execute highly complex research in an organized and self-driven manner. This experiment taught us what it means to work incredibly hard and to advocate for ourselves. We had to learn how to overcome roadblocks and keep working without obvious rewards. These lessons cannot be taught in any other high school setting. Beyond developing a detailed understanding of plant biology and gene editing, we have grown to understand the importance of setting and following extended timetables. No scientific research can occur without a driving vision and plan of execution, and we have grown to understand how to develop both of these elements throughout the year. These lessons will stay with us for the rest of our personal and professional careers, and we are very thankful for the opportunity to develop these skills. We have both been inspired to pursue scientific careers later in life. Martina Gilbert will study mining engineering, and Nathan Watervoort will study biochemistry.

Determining if *Caenorhabditis elegans* is a potential model organism for botulinum neurotoxin research

L. E. Rockwell and S. L. Fordham

Department of Science, Principals of Experimental Design in Biotechnology, Rock Canyon High School, Highlands Ranch, Colorado, USA

Botulinum neurotoxin has many medical uses and is currently heavily researched; however, no invertebrate has been identified as a potential model organism for botulinum neurotoxin research. Taking the first step in demonstrating whether *Caenorhabditis elegans* would be useful for future BoNT research could save researchers time and money and limit the use of vertebrates in early stages of research. In this research, deactivated BoNT/A tagged with Alexa 488 fluorescence were fed to *C. elegans*. The organisms were then washed and observed under a fluorescent microscope to confirm uptake of the neurotoxin. In a separate experiment, Botox, which contains BoNT/A, was also added to the *C. elegans* food source to observe the effects on general morphology and movement the toxin would have on the organism. After completing this research, no evidence was found to show that the toxin enters the *C. elegans* cells because the fluorescence levels from the wild type and the deactivated BoNT/A Alexa 488 were not significantly different. In addition, no evidence was found to show that *C. elegans* are physically affected by BoNT/A. Of the Botox exposed population, only 1.89% of the *C. elegans* were paralyzed, which can be attributed to natural variance and is not statically significant. The data collected in this research does not support the use of *C. elegans* as a useful model organism for BoNT research.

Botulinum neurotoxin (BoNT), the world's most potent toxin, is made by the anaerobic bacteria *Clostridium botulinum*. The bacteria, found in untreated water and soil, produces spores in low-oxygenated areas such as canned foods and contaminated honey. These spores produce BoNT. If ingested, the toxin can lead to botulism, a neurological disease resulting in flaccid muscle paralysis **(Pic. 1)**.[10]

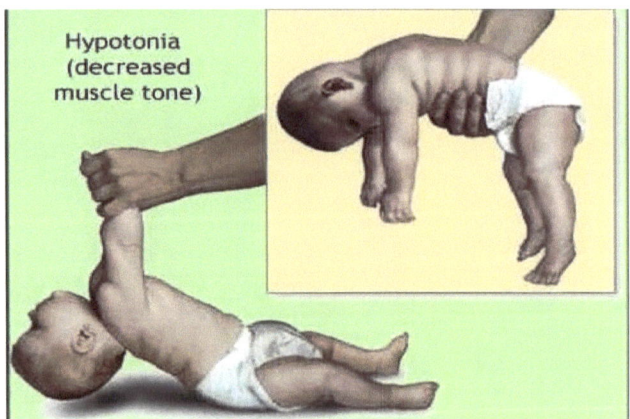

Picture 1: This image shows the effects of botulism to an infant's muscular system.[1]

Botulism disease blocks nerve transmission and has many symptoms including muscle weakness, muscle stiffness, difficulty swallowing or speaking, facial weakness, trouble breathing, and paralysis.[12] If left untreated, botulism may result in death. Treatment includes administrating an intravenous antitoxin, botulism immune globulin, and providing supportive care. As of 2014, botulism fatality rates were 5%-10%; the majority of deaths occurring in infants younger than 6 months old.[6] In most adults and children, natural defenses developed after six months of age degrade the bacteria before it produces the toxin.[7]

Botulism results from BoNT-mediated blockage of acetylcholine, a chemical in the body that functions as a neurotransmitter, and is released at nerve and muscle junctures.[5] There are seven serotypes of BoNT, from A-G, each produced by different strains of *C. botulinum*. BoNT/A is a single polypeptide chain cleaved by bacterial proteases at an exposed protein-sensitive loop, creating an active neurotoxin. Surrounding the neurotoxin is a complex of proteins called neurotoxin-associated proteins or NAPs. Because BoNT/A requires a high pH environment, nontoxic NAPs protect the neurotoxin from degrading in the gastrointestinal tract and allows for translocation across the intestinal mucosal layer. The NAPs then dissociate when the toxin is transferred from the gut into the bloodstream to the cells.[10] Once BoNT/A has entered the cell, it interacts with a family of proteins called SNARE proteins. These proteins facilitate the fusion of vesicles with their target membrane. BoNT/A interacting with these proteins, prevents synaptic vessels from fusing with the presynaptic membrane, resulting in the suppression of nerve impulses throughout the body. This can cause loss of muscle function and paralysis.

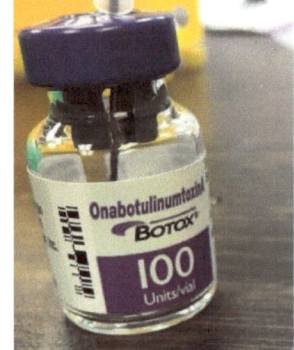

Picture 2: Image of Botox vial and type of Botox I used in this research.

BoNT/A is most commonly used to treat muscle control syndromes such as cervical dystonia, blepharospasm, facial nerve dysfunction, migraines, and is commercially used in Botox.[8]

Botox is a prescription medicine that works by injecting small amounts of BoNT/A directly into the muscle where it blocks the nerve transmission to that muscle (**Pic. 2**). The BoNT/A only affects the muscles and nerves it is injected; to and doesn't spread systemically. Because Botox contains minute amounts of BoNT/A, it is safe to use in our BSL-1 lab.

Mice, rats, and other small mammals are currently used in BoNT/A testing and research. For research studied, *C. botulinum* spores are directly injected into the infant test mice stomachs to give them botulism.[9] The mice and rats are then given the antitoxin in order to study how it works in the body. But using vertebrates in scientific research raises many ethical concerns. Using an invertebrate test subject that could be affected by BoNT would not only save time and money, but also limit the use of vertebrates in research.[11]

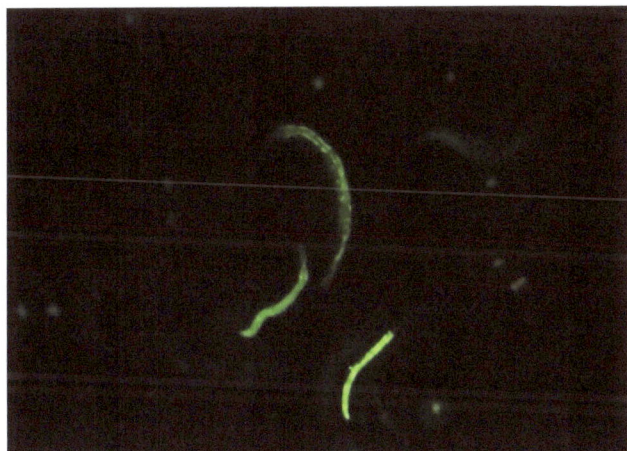

Picture 3: Photo of the auto fluorescence in wild type *C. elegans* taken by L. Rockwell.

Caenorhabditis elegans are one of the only invertebrates with a neuronal system (**Pic. 3**).[3] Also, they have the SNAP-25 proteins that can be inflicted with BoNT/A.[13] Recent research showed evidence that *C. elegans* are not affected by BoNT/A. In their research, adult *C. elegans* exposed to the toxin did not show signs of paralysis within 24 hours.[2] This research did not identify if the toxin is taken into the cells of *C. elegans* or whether the organism can be used in BoNT/A research. More testing is needed to determine whether the organism can resist the toxin or if the toxin even enters the organism through uptake into their cells.[1] Despite their results, it is important identify whether it is taken into its cells and confirm whether BoNT/A could potentially affect the organism. I used a fluorescently tagged deactivated version of the BoNT/A protein, called Dr.BoNT/A, and observed whether it enters into the *C. elegans*. Dr.BoNT/A has all the same properties as BoNT/A but lacks the ability to turn off the SNAP-25 and SNARE proteins, making it unable to inflict harmful neurological effects. The Dr.BoNT/A was tagged with Alexa 488 fluorescent dye to show whether the *C. elegans* would uptake the protein through feeding. Last, in order to confirm the results from the previous toxicity study, I also tested the effect BoNT/A has on the organism by adding Botox solution to their food source, and then measuring the subsequent rate of paralysis. This research is a critical step in identifying the use of *C. elegans* as a potential model organism for future BoNT/A research.

METHODS
Fluorescence Pre-Trials

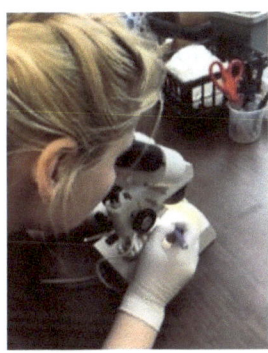

Picture 4: Image of Rockwell observing N2 Wild Type *C. elegans*.

During pre-trials, I tested a 1uM concentration of BSA proteins and BSA tagged with Alexa 488 fluorescence to determine if there is sufficient uptake into the *C. elegans* to visualize above its natural fluorescence. A measurable difference in the levels of fluorescence using Alexa 488 would show whether the target protein entered the *C. elegans*.

I received pure BSA protein tagged with Alexa 488, which naturally fluoresces at 488nM, from Dr. Bal Ram Singh, the director of the National Botulinum Research Center. 100mm NGM agar petri plates obtained from IPM Scientific were seeded with 100uL of a 1uM Alexa 488 tagged BSA/LB broth/OP50 solution (**Pic. 4**). The OP50 bacteria were cultured overnight at 37°C in the LB broth, and then the Alexa 488 BSA solution was added (**Pic. 5**). As a control, I also seeded 100uL of a 1uM non-fluorescently labeled BSA/LB broth/OP50/ solution on to a second 100mm NGM agar plate.

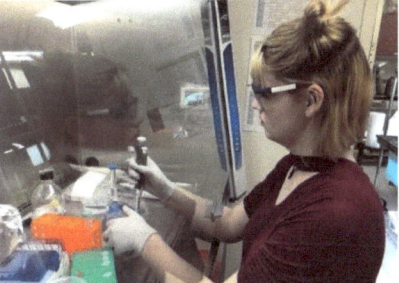

Picture 5: Image of Rockwell making the BSA solutions.

These plates were dried in the biological safety cabinet for one hour, after which the N2 *C. elegans* were then chunked onto each plate to get a variety of life stages exposed to the proteins onto the plates. The plates were then shipped to the Botulinum Research Center (BRC) at the Institute of Advanced Sciences in Boston where Dr. Raj Kumar imaged the *C. elegans* with the high powered fluorescent microscope (**Pics. 6 and 7**). **Picture 6** shows the images taken of the *C. elegans* form the BSA non-fluorescing control plate. **Picture 7** shows the images taken of the *C. elegans* from the BSA/Alexa 488 experimental plate.

In the experimental images (**Pic. 7**) two out of the five *C. elegans* demonstrated a higher amount of fluorescence when compared to the control *C. elegans* (**Pic. 6**). As seen in **Picture 7** below, image 1 and 4 appeared to have a higher fluorescence as compared to the images in **Picture 6** (**Pics. 6 and** 7). Due to this data, my mentors felt that I should move forward with the experiment.

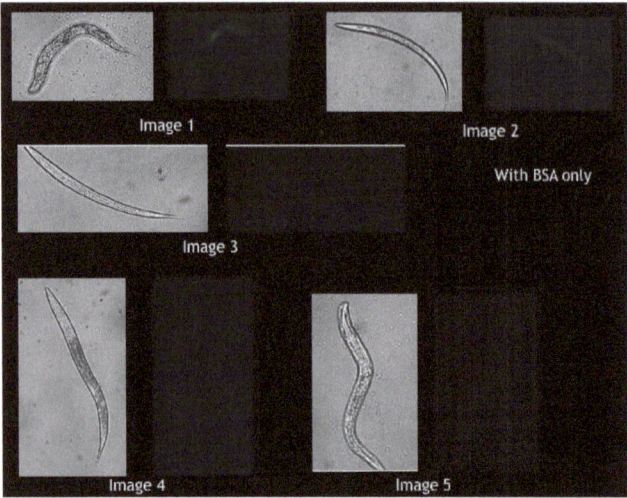

Picture 6: Images of five *C. elegans* picked off the control plate with the BSA non-fluorescing proteins.

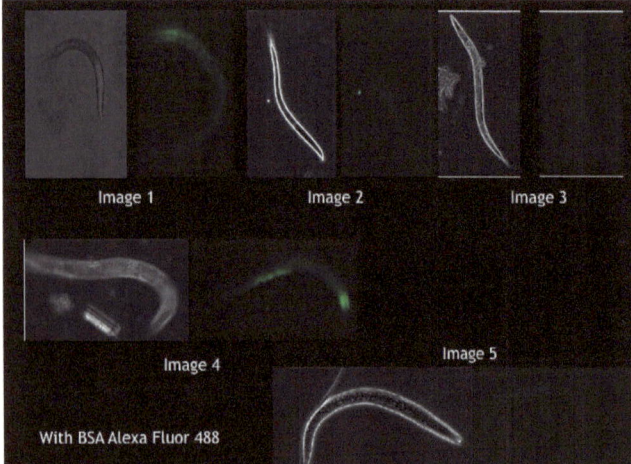

Picture 7: Images of five *C. elegans* picked off the control plate with the BSA/Alexa 488 proteins

In this trial Dr.BoNT/A, proteins tagged with Alexa 488 fluorescence were fed to *C. elegans* to observe if the proteins can be taken into the organism through feeding. A 100mm NGM agar plate was seeded with 150uL of a 1uM Dr.BoNT/A/Alexa 488/LB broth/ solution using the same protocols as in the fluorescence pre-trials. A control plate was also made with a 1uM non-labeled Dr.BoNT/A/LB broth/OP50. After the *C. elegans* were chunked onto the plates, they were also sent to the BRC to be observed.

For this part of the experiment, I exposed the *C. elegans* to three different concentrations of Botox and measured the rate of subsequent paralysis. The Botox solution was diluted with the LB broth/OP50 culture to create the three different dilutions- 1:10, 3:10, and 1:2- used in this

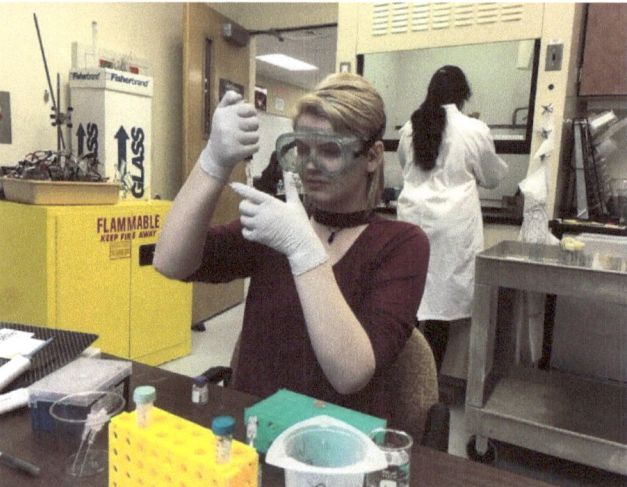

Picture 8: Photo of Rockwell preparing the Botox solutions.

experiment (**Pic. 8**). I also used a LB broth/OP50 culture without the toxin as a control.

A 100/mm Y divided NGM petri plate was seeded with each Botox dilution by adding 100uL to each well. After the plates dried, I picked three adult worms onto each well and left them overnight in an incufridge at 16ºC to lay eggs. The next day, I picked off all adult worms, leaving only eggs and L1s in order to create a synchronized population.

I examined the *C. elegans* at 24 hour intervals and picked off the dead or fully paralyzed *C. elegans*. Dead *C. elegans* were identified by having no response to touch with a platinum wire pick, while fully paralyzed *C. elegans* only move their heads.

RESULTS
Dr.BoNT/A Trials

After exposing the *C. elegans* to the Dr.BoNT/A Alexa 488 solution, they were imaged using a high powered fluorescent microscope (**Pics. 9 and 10**). As seen in the images below, through visual confirmation, the *C. elegans* exposed to

Dr.BoNT/A/Alexa 488 fluorescence did not seem to show an obvious difference in the fluorescence levels compared to the non-labeled Dr.BoNT control (**Pics. 9 and 10**). In the Dr.BoNT/A control, the organs of the *C. elegans* seem to be fluorescing more than those exposed to the Dr.BoNT/A/Alexa 488. This could be attributed to background fluorescence from the Alexa 488 in the media of the treatment plate infused with the fluorescence protein solution (**Pics. 9 and 10**).

To confirm further that there was no difference in the fluorescence between the treatment and control, the mean gray value was calculated from the images (**Graph 1**). The mean gray value is the sum of the gray values in each pixel

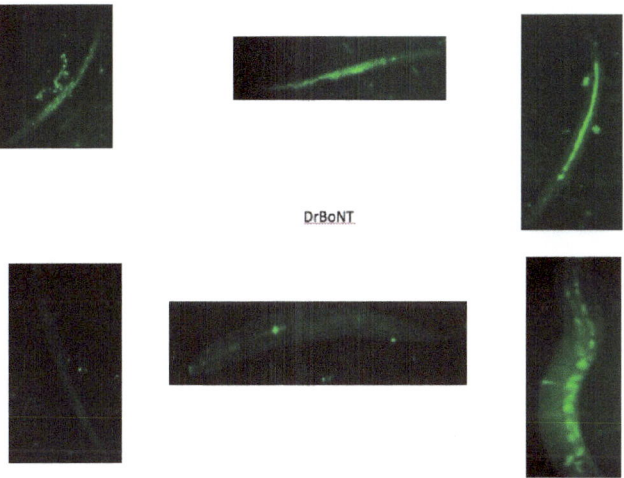

Picture 9: Images of *C. elegans* from the control plate with the 1 uM Dr.BoNT/A/Alexa 488 fluorescence.

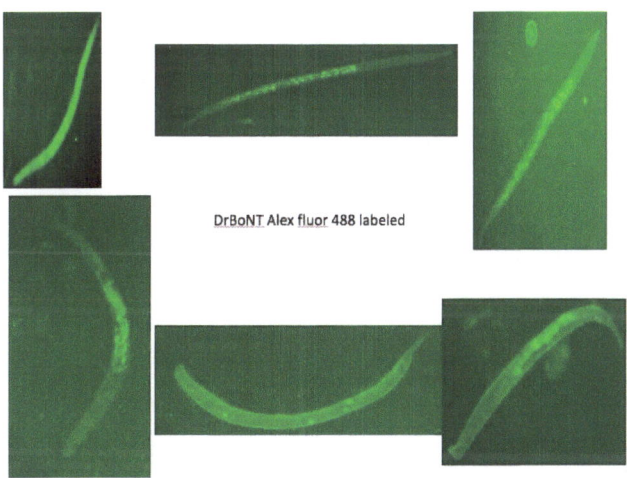

Picture 10: Images of *C. elegans* from the control plate with the 1 uM non-labeled Dr.BoNT/A. This control shows the natural levels of fluorescence in *C. elegans*.

of an image and is used to calculate the levels of fluorescence. Statistical analysis showed no significant difference in the mean gray values between the fluorescence in the control and the Dr.BoNT/A/Alexa 488 treated *C. elegans*. To analyze the data further, an ANOVA was performed. The p-value of 0.316 > 0.05, showed no statically significant difference in the fluorescence between the *C. elegans* on the control plate and the *C. elegans* on the experimental plate.

Botox Trials

After conducting the three trials, I found no evidence supporting Botox affects the *C. elegans*. From the 1:10 and the 3:10 treatments, no *C. elegans* showed obvious phenotypic changes and 100% of the population remained unaffected throughout a seven-day period (**Graph 2**). In the

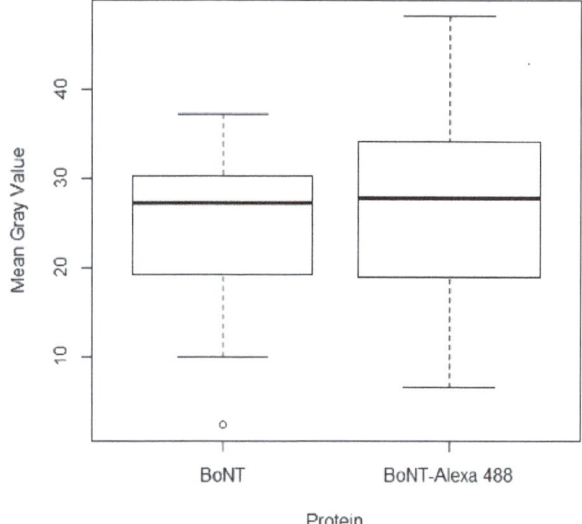

Graph 1: Boxplot showing the mean gray value of the fluorescence between the control and experimental *C. elegans*.

1:2 Botox treatment, only one *C. elegans* became paralyzed, which was a mere 1.89% of the total population (**Graph 2**).

Using a 95% confidence interval for the 1:2 Botox concentration, the p-value was 0.0187 with a 95% C[-0.0178, 0.0555], which demonstrates that this single paralyzed *C. elegans* can be attributed to natural variation in the population and not attributed to the Botox.

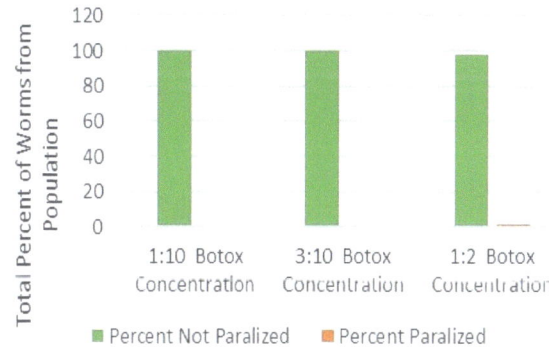

Graph 2: This graph represents the percentage of paralyzed and non-paralyzed *C. elegans* out of the whole population for each Botox treatment at the end of seven days.

DISCUSSION

Botulinum neurotoxin, the world's most potent toxin, is extensively used in the medical field. Discovering if *C. elegans* can be used as a model organism for BoNT research would not only save researchers time and money, but also avoids ethical issues that come from using vertebrates in research. Using a lower order, simpler organism could also help researchers discover a new way to screen and facilitate BoNT detection and discover more about BoNT's action mechanism. Previous research showed evidence that *C. elegans* cannot be affected by BoNT/A.[2] However, it was never confirmed whether the toxin was actually taken into the cells of *C. elegans* or if its neurons simply were not affected.

To test this, I exposed the *C. elegans* to a 1uM concentration of deactivated BoNT/A tagged with Alexa 488 fluorescence solution in order to visualize the amount of uptake. The mean gray values of the images taken of the *C. elegans* exposed to Dr.BoNT/A/Alexa 488 fluorescence were not statically different than the mean gray value of the images taken of the *C. elegans* on the control plate and those exposed to the non-fluorescing Dr.BoNT/A on the experimental plate. This data provided evidence that there was little to no uptake of Dr.BoNT/A into the organism through feeding.

When exposing the *C. elegans* to varying concentrations of Botox, containing active BoNT/A protein, only 1.89% of the *C. elegans* in the 1:2 Botox treatment showed signs of paralysis. All other dilutions resulted in 0% paralysis in the *C. elegans*. This 1.89% was statistically insignificant and confirms the previous research that showed BoNT/A does not result in paralysis of *C. elegans*.

It appears that some targeted proteins, including the Alexa 488 fluorescence, have the ability to get into the *C. elegans* through the cuticle or by feeding as shown in pre-trials. However, the Dr.BoNT/A/Alexa 488 fluorescence and Botox trials there was no evidence that the BoNT/A protein either enters or affects the organism. In order to identify whether BoNT could affect the neuronal cells in the *C. elegans*, a future study could attempt using a microinjection technique to enable the proteins to enter the *C. elegans*.

Microinjection is a commonly used method in *C. elegans* research to inject substances, such as DNA or RNA, directly into the cells.[4] If BoNT were directly injected into the cells of *C. elegans*, the experiment could show whether the BoNT has an effect on *C. elegans*. Alternatively, *C. elegans* neuronal cells could be cultured and exposed to the BoNT and studied directly in culture. But identifying the actual reasons why the proteins do not affect the *C. elegans* and is beyond the scope of this research.

One error I encountered was contamination on the Botox NGM plates. There was a considerable amount of contamination due to the fact the Botox was not completely sterile. The excess bacteria may have altered my results in some unforeseen way. This is unlikely, however, given the *C. elegans* remained unparalyzed throughout the experiment. Another challenge I encountered was the discovery that *C. elegans* has a natural green auto fluorescence, and using green fluorescent protein to visualize the presence of the protein was not the best way to approach this research. If this research is repeated, the BoNT should be tagged with a different fluorescent molecule, such as a RFP. In addition, a very small population of the *C. elegans* were imaged in the Dr.BoNT/A experiment. Because of the small sample size

and small quantity of reagent available to me, I was only able to test a 1uM concentration one time and was unable to repeat these trials in order to collect more data and verify my results.

ACKNOWLEDGMENTS
I would like to thank Dr. Bal Ram Singh, Dr. Raj Kumar, and Dr. Soniya Balli of the Botulinum Research Center for their mentorship, guidance, and for providing me with the Dr. BoNT/A used in this research. Also, I am grateful to Dr. Angelina Kohler teaching me about Botox. I want to express my gratitude to Janet Rockwell and Whitney Lord for donating funds for my research. I would also like to thank Susanne Petri and Nikki Dobos for sharing their lab space and for supporting me throughout this research; Wendy Lerolland and Bryan Winkelman for helping me find resources, making edits on my article, and controlling the funding and website; Shreemathi Harikrishnan and Hannah Phillip for helping me perform some of my methods, and Nathan Watervoort and Tina Gilbert for giving me their advice and technical support throughout my research. Lastly, I would like to thank Rock Canyon High School and Douglas County School District for providing the facilities and equipment needed for this research. Without all of you this research would not have been possible. Thank you.

REFERENCES
1. Beatty, N. J. (2016). Clostridium botulinum: honey and home-canned foods. Pediatric Infectious Disease. Retrieved, 2017, March 11. [Web]
2. Chang, T. W. & Lin, K. H. Effect of *Clostridium botulinum* type A neurotoxin (BoNT/A & TANC) on *C. elegans*.
3. Emmons, S.W. (2005). Male development WormBook, The *C. elegans* Research Community, WormBook, DOI:10.1895/wormbook.1.33.1
4. Evans, T. C. (2006). Transformation and microinjection. Wormbook. Retrieved 2017, April 04. [Web]
5. Fischer, A., Nakai, Y., Eubanks, L. M., Clancy, C. M., Tepp, W. H., Pellett, S., & Montal, M. (2009). Bimodal modulation of the botulinum neurotoxin protein-conducting channel. *Proceedings of the National Academy of Sciences, 106*(5), 1330-1335.
6. Gill, R. L. (2015). Infant Botulism. (2015) Retrieved 2016, October 12. [Web]
7. Mandal, M. D. (2014). Botulism prognosis news *medical life sciences*. Retrieved 2017, March 12. [Web]
8. Ney, J. P., & Joseph, K. R. (2007). Neurologic uses of botulinum neurotoxin type A. *Neuropsychiatric Disease and Treatment, 3*(6), 785–798.
9. Pellett, S. Tepp, W. H. Scherf, J. M. & Johnson, E. A. (2015). Botulinum neurotoxins can enter cultured neurons independent of synaptic vesicle recycling. *PloS one, 10*(7), e0133737.
10. Rosales, R. L., Bigalke, H., & Dressler, D. (2006). Pharmacology of botulinum toxin: Differences between type A preparations. *European Journal of Neurology, 13*(s1), 2-10.
11. Singh, B. R. & Kukreja, R. (2015). The botulinum toxin as a therapeutic agent: Molecular and pharmacological insights. *Research and Reports in Biochemistry*. DOI:10.2147/rrbc.s60432.
12. Sugiyama, H. (1979). Animal models for the study of infant botulism. *Review of Infectious Diseases, 1*(4), 683-688.
13. Sulston J, & Hodgkin J. (1988). Methods. The nematode *Caenorhabditis elegans* (Wood WB, ed). Cold Spring Harbor, NY. [Print]

ABOUT THE AUTHOR

Pictured: Leilani Rockwell and her mentor Dr. Bal Ram Singh. Not pictured Dr. Raj Kumar and Dr. Soniya Balli.

Through this research, I have learned many things. Besides learning new lab skills and protocols I have learned how to ask for help. I have also learned how to keep going and trying after continuous failing. The biggest thing I learned is how to fight. Living with chronic migraines makes it hard to do anything, but this course made me fight through my pain no matter what. Another thing I learned is that before starting research, I should make sure I have all the information necessary. During this research, I found that the *C. elegans* auto fluoresce green-meaning I shouldn't have used the green Alexa fluorescence, but found a different color.

After graduating I am going to study marine biology and Korean language. I will take these skills I learned with me for the rest of my life.

5810 McArthur Ranch Road
Highlands Ranch, CO 80124
303-387-3000

Principal
Andy Abner
Andrew.Abner@dcsdk12.org

Registrar
Polly Poindexter
Polly.Poindexter@dcsdk12.org

Administrative Assistant
Barb Cocetti
Barbara.Cocetti@dcsdk12.org

STEM PROGRAMMING

The Principals of Experimental Design in Biotechnology course is one of many courses offered as part of our choice-driven STEM programming, which allows each of our students to prepare for their vision of a career in science, technology, engineering, or math.

Due to the competitive nature of STEM majors in college, we believe that taking a rigorous course load, including Honors, AP, and dual credit courses, is the best way to prepare students for the coursework they will encounter. In addition, involvement in clubs that encourage competition in Science, Technology, Engineering, and Business allows students the opportunities to think on their feet, construct and communicate arguments, and work through the engineering process. Finally, our wish is that students will become involved in an internship or shadowing experiences in order to gain the workplace experience that our classes may not provide.

Students who diligently pursue this difficult course load, as well as meeting these additional requirements, will not only benefit from their knowledge and preparation, but will also be able to show the universities that they are determined students by presenting them with a STEM certificate.

Rock Canyon High School
Home of the Jaguars

Our Mission:
To Empower, To Explore, To Encourage and To Excel in Education

Our Vision:
Our student-centered culture practices collaborative decision making and continuous improvement in a safe, supportive environment.

Rock Canyon is a comprehensive high school consisting of grades nine through twelve, located in Highlands Ranch, Colorado, a southern suburb of Denver. Our community is composed primarily of working professionals.

Rock Canyon is part of the Douglas County School District (DCSD), the third largest school district in Colorado, serving over 67,000 students for the 2016-2017 school year. The district is comprised of 9 high schools, 9 middle schools, 47 elementary schools, 12 charter schools, 2 magnet schools, 3 alternative schools and an online school. The DCSD continues to maintain its standing as one of the finest, highest achieving districts in Colorado.

Rock Canyon opened in 2003. It has a current enrollment of 2,180 students. RCHS occupies a 279,250 square foot building on an 80-acre campus.

Rock Canyon High School prides itself on excelling in academics, activities and athletics to create a balanced and comprehensive high school experience for all students. We strive to develop a tradition of excellence in order to develop a premier high school program for all post-secondary options. Rock Canyon is currently ranked as one of the top high schools in Colorado.

We invite our parents to take an active role in their student's education by empowering their students to explore the many opportunities offered at Rock Canyon while continuing to encourage their students to excel in their educational goals. We truly believe a partnership must exist between the school and the family; together we can elevate our students to the next level.